IRISH RIVERS:
BIOLOGY AND MANAGEMENT

Proceedings of a seminar held on 23 and 24 February 1989

edited by

Martin Steer

Department of Botany, University College Dublin

ROYAL IRISH ACADEMY
DUBLIN 1991

ROYAL IRISH ACADEMY

19 Dawson Street

Dublin 2

Copyright ©1991 Royal Irish Academy

ISBN 0 901714 90 9

British Library Cataloguing-in-Publication Data.

A catalogue record for this book is available from the British Library

Also published by the

Royal Irish Academy

Renewable Resource Development in Ireland

edited by A.C. Cassells

Biological Indicators of Pollution

edited by D.H.S. Richardson

Taxonomy : Putting Plants and Animals in their Place

edited by Christopher Moriarty

Ecology and Convservation of Irish Peatlands

edited by Gerry Doyle

CONTENTS

ORGANISING COMMITTEE

The organising committee for the seminar included nominees of the Royal Irish Academy's National Committee for Biology and representatives of the Central Fisheries Board, Glasnevin.

NOMINATED BY THE NATIONAL COMMITTEE FOR BIOLOGY

PROFESSOR J. KAVANAGH
Department of Crop Science, Horticulture and Forestry, University College Dublin

DR CHRIS MORIARTY
Fisheries Division, Department of the Marine, Leeson Lane, Dublin 2

PROFESSOR MARTIN STEER
Department of Botany, University College Dublin

REPRESENTATIVES OF THE CENTRAL FISHERIES BOARD

DR PADDY FITZMAURICE
Central Fisheries Board, Balnagowan, Mobhi Boreen, Glasnevin, Dublin 9

DR KEN WHELAN
Salmon Research Agency, Newport, Co. Mayo

LIST OF CONTRIBUTORS

DR JOHN BRACKEN
Department of Zoology, University College Dublin

MR JOSEPH CAFFREY
Central Fisheries Board, Glasnevin, and Department of Botany, University College Dublin

DR RICHARD CRESSWELL
National Rivers Authority, Welsh Region, River House, St Mellons, Cardiff CF3 0LT, Wales

DR WALTER CROZIER
Department of Agriculture for Northern Ireland, Fisheries Research Laboratory, Coleraine, Co. Londonderry

PROFESSOR JIM DOOGE
President of the Royal Irish Academy; Emeritus Professor of Civil Engineering, University College Dublin

DR CHARLES ESSERY
Department of Computer Science, University of Ulster at Jordanstown, Shore Road, Newtownabbey BT 37 0QB, Co. Antrim

DR PADDY FITZMAURICE
Central Fisheries Board, Glasnevin, Dublin 9

DR PAUL GILLER
Department of Zoology, University College Cork

DR MARY KELLY-QUINN
Department of Zoology, Trinity College Dublin

DR GERSHAM KENNEDY
Department of Agriculture for Northern Ireland, Fisheries Research Laboratory, Coleraine, Co. Londonderry

MR JOHN LUCEY
Environmental Research Unit, Waterloo Road, Dublin 4

MR SEAN McMORROW
Barrister-at-Law, 40 Hollybrook Road, Dublin 3

MS MARY NORTON
Department of Zoology, University College Dublin

DR MARTIN O'FARRELL
Department of Zoology, Trinity College Dublin

DR MARTIN O'GRADY
Central Fisheries Board, Glasnevin, Dublin 9

DR STEPHEN ORMEROD
Catchment Research Group, School of Pure and Applied Biology, University of Wales, College of Cardiff, c/o NRA (Wales), Penyfai House, Furnace, Llanelli, Dyfed SA 15 4EL, Wales

MS HELEN PHILLIPS
Department of Zoology, University College Dublin

MR MARTIN QUINN
Electricity Supply Board, Ardnacrusha, Co. Clare

MR GRAHAM RUTT
Catchment Research Group, School of Pure and Applied Biology, University of Wales, College of Cardiff, c/o NRA (Wales), Penyfai House, Furnace, Llanelli, Dyfed SA 15 4EL, Wales

DR PAUL TONER
Environmental Research Unit, Pottery Road, Dun Laoghaire, Co. Dublin

DR EILEEN TWOMEY
Fisheries Research Centre, Castleknock, Dublin 15

DR HELENA TWOMEY
Department of Zoology, University College Cork

DR KEN WADE
Welsh Water, Bridgend, West Glamorgan, Wales

MR NEILL WEATHERLEY
Catchment Research Group, School of Pure and Applied Biology, University of Wales, College of Cardiff, c/o NRA (Wales), Penyfai House, Furnace, Llanelli, Dyfed SA 15 4EL, Wales

PROFESSOR BRENDAN WHELAN
Economic and Social Research Institute, 4 Burlington Road, Dublin 4

DR KEN WHELAN
Salmon Research Agency, Newport, Co. Mayo

PROFESSOR DAVID WILCOCK
Department of Environmental Studies, University of Ulster at Coleraine, Cromore Road, Coleraine BT52 1SA, Co. Londonderry

LIST OF PARTICIPANTS

Mr J. Allison
Environmental Sciences Unit, Trinity College Dublin

Mr A.P. Barry
South Western Regional Fisheries Board

Mr R. Barry
Duphar Ireland Ltd

Ms M.M. Berigan
Department of Zoology, University College Dublin

Mr C. Blackwell
Department of Materials Engineering and Industrial Chemistry, N.I.H.E., Limerick

Mr P. Bourke
Eastern Regional Fisheries Board

Mr O. Boyle
Department of Environment

Dr P. Bradley
Royal Society for the Protection of Birds, Belfast

Mr M. Brady
Environmental Services, Navan

Dr J. Bracken
Department of Zoology, University College Dublin

Mr J. Brown
Fisheries Research Centre, Dublin

Mr C. Byrne
Department of Zoology, University College Dublin

Mr L. Byrne
Bord na Móna

Mr J. Caffrey
Central Fisheries Board and University College Dublin

Ms R. Callaghan
Department of Zoology, University College Galway

Mr L. Carley
Rockford College, Illinois, U.S.A.

Mr J. Clancy
Shannon/Elbe Study, EOLAS

Ms F. Clarke
Southern Regional Fisheries Board

Mr S. Clinton
Limerick County Council

Mr J. Coady
Environmental Sciences Unit, Trinity College Dublin

Mr E.P. Conway
Irish Planning Institute

Dr D.C.F. Cotton
School of Science, Regional Technical College, Sligo

Ms G. Country
Department of Materials Engineering and Industrial Chemistry, N.I.H.E., Limerick

Ms S.M. Coyle
Department of Zoology, University College Dublin

Mr G. Crawford
Foyle Fisheries

Dr R. Cresswell
Welsh Water Authority

Dr W. Crozier
Department of Agriculture for N.I.

Mr A. Cullagh
Southern Regional Fisheries Board

Mr L. Curran
Shannon/Elbe Study, EOLAS

Mr J. Curtin
Engineering Branch, O.P.W.

Mr E. Cusack
Eastern Regional Fisheries Board

W.G. Dallas
Enviroplan Services Ltd, Kells

Mr B.T. Daly
Department of Agriculture

Mr P. Diggin
Department of Materials Engineering and Industrial Chemistry, N.I.H.E., Limerick

Mr I. Dinan
Department of Materials Engineering and Industrial Chemistry, N.I.H.E., Limerick

Mr A. Doherty
Clane Trout and Salmon Anglers Association

Prof. J.C.I. Dooge, P.R.I.A.
Centre for Water Resources Research, University College Dublin

Ms C. Douglas
Wildlife Service, O.P.W.

Dr C.I. Essery
Department of Computer Science, University of Ulster at Jordanstown

Dr J.S. Faulkner
Department of the Environment for N.I.

Dr P. Fitzmaurice
Central Fisheries Board

Mr M. Fitzsimons
Shannon Regional Fisheries Board

Mr R. Fluskey
Fisheries Research Centre, Dublin

Dr G. Forde
Lough Inagh Fishery, Co. Galway

Ms E. Frost
Department of Materials Engineering and Industrial Chemistry, N.I.H.E., Limerick

Dr J.J. Gardiner
Department of Forestry, University College Dublin

Mr P. Gargan
Central Fisheries Board

Dr P. Giller
Department of Zoology, University College Cork

Mr M. Gillooly
Environmental Sciences Unit, Trinity College Dublin

Mr R. Goodwillie
Wildlife Service, O.P.W.

Dr H. Gracey
Department of Agriculture for N.I.

Dr N.F. Gray
Department of Microbiology, Trinity College Dublin

Mr N. Hackett
South Western Regional Fisheries Board

Mr M. Hallinan
Faculty of Agriculture, University College Dublin

Ms G. Hannigan
Eastern Regional Fisheries Board

Mr J. Harte
Limerick Corporation

Ms D. Harvey
Department of Materials Engineering and Industrial Chemistry, N.I.H.E., Limerick

Mr V. Hayes
Kerry County Council

Mr R. Hernan
Department of Zoology, University College Dublin

Dr H. Heuff
Blackrock, Co. Dublin

Mr M. Holligan
Kildare County Council

Mr W. Johnston
Training Division, O.P.W.

Prof. J. Kavanagh
Department of Plant Pathology, University College Dublin

Dr M. Kelly-Quinn
Department of Zoology, Trinity College Dublin

Dr G. Kennedy
Department of Agriculture for N.I.

Mr T.K. Kennedy
Department of Zoology, University College Dublin

Mr P. Kilfeather
Southern Regional Fisheries Board

Mr J. Lucey
Environmental Research Unit, Kilkenny

Ms J. Lynch
Department of Zoology, University College Dublin

Dr G. McCall
Stradbally, Co. Laois

Mr E. McCann
Department of Materials Engineering and Industrial Chemistry, N.I.H.E., Limerick

Mr C. McCarthy
Department of Materials Engineering and Industrial Chemistry, N.I.H.E., Limerick

Dr R. McCarthy
Coillte Teoranta, Bray, Co. Wicklow

Dr T.K. McCarthy
Department of Zoology, University College Galway

Mr J.F. McCormack
Environment Services, Navan

Mr W. McCumiskey
Environmental Research Unit

Ms K. McGibney
Department of Zoology, University College Dublin

Mr J. McInerney
Engineering Branch, O.P.W.

Ms E. McKechnie
Department of Zoology, University College Cork

Ms K. McLaughlin
Environmental Sciences Unit, Trinity College Dublin

Ms C. McLoughlin
Eastern Regional Fisheries Board

Mr R. McLoughlin
Environmental Sciences Unit, Trinity College Dublin

Mr S. McMorrow
Barrister-at-Law, Four Courts

Mr D. McNamara
Department of Materials Engineering and Industrial Chemistry, N.I.H.E., Limerick

Mr S. Mansfield
Environmental Sciences Unit, Trinity College Dublin

Ms A. Martin
Cell Culture Laboratory, N.I.H.E., Dublin

Mr S. Monahan
Boyne Research Centre

Mr A. Morgan
Engineering Branch, O.P.W.

Dr C. Moriarty
Fisheries Research Centre, Dublin

Mr M. Murphy
Dept of Zoology, University College Galway

Dr. D. Murray
Department of Zoology, University College Dublin

Mr D. O'Brien
Environmental Sciences Unit, Trinity College Dublin

Dr D. O'Brien
Earthwatch

Mr J. O'Brien
Eastern Regional Fisheries Board

Dr N. O'Carroll
Forest Service, Department of Energy

Ms J. O'Connell
Department of Zoology, University College Cork

Ms P. O'Connor
South Western Regional Fisheries Board

Dr M. O'Farrell
Department of Zoology, Trinity College Dublin

Dr M. O'Grady
Central Fisheries Board

Mr S. O'Halloran
Department of Materials Engineering and Industrial Chemistry, N.I.H.E., Limerick

Mr L. O'Keeffe
Rintoul, Young and McMordue, Belfast

Mr N. O'Maoileidigh
Department of Zoology, University College Dublin

Mr J. O'Neill
Shannon/Elbe Study, EOLAS

Mr J. O'Regan
Engineering Branch, O.P.W.

Mr P. O'Sullivan
Department of the Marine

Dr S. Ormerod
Department of Pure and Applied Biology, University of Wales, Cardiff

Mr A. Petersen
Regional Technical College, Cork

Ms H. Philips
Department of Zoology, University College Dublin

Mr R. Poole
Department of Zoology, Trinity College Dublin

Mr M. Quinn
Electricity Supply Board, Co. Clare

Dr P.J. Raven
Department of the Environment for N.I.

Ms J. Reynolds
Clare County Council

Mr J.B. Rogers
Southern Regional Fisheries Board

Mr K. Rogers
Western Regional Fisheries Board

Mr J. Ryan
Wildlife Service, O.P.W.

Miss M.J.P. Scannell
National Botanic Gardens

Ms S. Sheil
North Western Regional Fisheries Board

Ms C. Smith
Clare County Council

Dr S. Smith
Freshwater Biology Investigation Unit, Queen's University, Belfast

Mr J. Spillane
Shannon Regional Fisheries Board

Ms M. Stack
Cork County Council

Mr J. Stapleton
Eastern Regional Fisheries Board

Professor M. Steer
Department of Botany, University College Dublin

Mr J. Stewart
I.C.I. Ireland Ltd

Mr T. Sullivan
Southern Regional Fisheries Board

Mr J. Taggart
Department of Biology, Queen's University, Belfast

Dr P. Timpson
Sligo Regional Technical College

Ms C. Tobin
Freshwater Laboratory, University of Ulster, Traad Point, Magherafelt, Co. Derry

Dr P. Toner
Environmental Research Unit

Miss E. Twomey
Fisheries Research Centre

Dr Helena Twomey
Department of Zoology, University College Cork

Mr T. Walker
Environmental Sciences Unit, Trinity College Dublin

Mr L. Walsh
Tara Mines Ltd, Navan

Dr N. Weatherley
University of Wales, Cardiff

Professor B. Whelan
Economic and Social Research Institute

Dr K.F. Whelan
Central Fisheries Board

Dr K. Whitaker
Salmon Research Trust of Ireland

Professor D.N. Wilcock
Department of Environmental Studies, University of Ulster at Coleraine

Ms H. Witts
Science Library, University College Dublin

Mr S. Wolfe-Murphy
Freshwater Biology Investigation Unit, Queen's University, Belfast

ACKNOWLEDGEMENTS

The editor wishes to thank all those who assisted with the organisation of the seminar and the publication of this volume. The topic was selected by the Royal Irish Academy National Committee for Biology, the body responsible for organising this series of annual seminars. The work of seminar organisation was effectively undertaken by a committee composed of Dr Paddy Fitzmaurice and Dr Ken Whelan (both of the Central Fisheries Board at the time) and Professor Jim Kavanagh, Dr Chris Moriarty and Professor Martin Steer of the National Committee for Biology.

Generous sponsorship of the meeting was provided by the Department of the Environment, the Electricity Supply Board and many others, listed below.

The text of this publication was typeset in the Department of Botany, University College Dublin, by Ms Jane Flaherty and Ms Grainne Murphy with the generous assistance of Dr Gerry Doyle. Valuable editorial assistance was given by Ms Barbara Young and the staff of the publications office, Royal Irish Academy.

We are all indebted to Ms Sara Whelan, Royal Irish Academy, for shepherding the various committees involved in this enterprise and undertaking the organisation of the seminar within the Academy.

MAJOR SPONSORS

Department of the Environment

Electricity Supply Board

OTHER SPONSORS

B & I Line

Bord Fáilte

P.J. Carroll & Co. plc

Central Fisheries Board

Department of the Marine

Duphar Ireland Ltd

ICI Ireland Ltd

Irish Salmon Growers' Association

The National Development Corporation Ltd

Professor Tom Raftery, M.E.P.

The Salmon Research Trust of Ireland Inc.

Tara Mines Ltd

Trinity College Dublin, Department of Zoology

University College Dublin, Department of Botany

University College Dublin, Department of Zoology

University College Galway, Department of Botany

University of Ulster Freshwater Laboratory

Wildlife Service, Office of Public Works

Part 1

Introduction

In: Steer, M.W. (ed.) 1991 *Irish Rivers : Biology and Management*, p. 3. Royal Irish Academy, Dublin.

INTRODUCTION

Paddy Fitzmaurice

Central Fisheries Board, Dublin

The estimated length of main river channels in Ireland is greater than 13,500 km. Consequently, the country is well endowed with a valuable, self-renewing natural resource. Rivers are put to many uses and are exploited by numerous sectional interests for specific purposes. Often these sectional interests appear to operate in isolation and may not fully take account of other beneficial uses or impacts on the environment.

The National Committee for Biology of the Royal Irish Academy, being aware of the incalculable value and multi-purpose role of rivers and knowing that proper rational management of this resource can only benefit the country, decided to hold a seminar in the Royal Irish Academy on 23rd and 24th February, 1989. This was the first national seminar dealing with the biology and management of Irish rivers.

Rivers are dynamic systems with constantly changing characteristics. The quality of the water, the form, gradient and flow (etc.) in the channel are all expressed in the biological communities present. Any organism, species or population is incapable of existing without its proper environment. Therefore there is an interaction between biological communities and their environment. Any changes in the environment will affect the biological components of the ecosystem. The extent of the changes will depend on the adaptive powers of the biological species present. Therefore there is a necessity to manage the ecosystem in a rational way to ensure that all interests are catered for and none suffer irreparable loss.

The Committee were aware that the Department of the Environment has laid great emphasis on water pollution control and on water quality in our river systems and that the fisheries interests were being looked after by the Department of the Marine and the Fisheries Boards. Other government departments and agencies were responsible for areas such as arterial drainage and hydropower, which impact on river ecosystems. With so many government agencies involved, it could be stated that the management of rivers is generally a state responsibility.

The following chapters illustrate various aspects of the interaction between biology and management of Irish rivers. These form a broad base which will hopefully be of value in the formulation of fully coordinated management plans.

In: Steer, M.W. (ed.) 1991 *Irish Rivers : Biology and Management,* pp 5-26.
Royal Irish Academy, Dublin.

THE FLOW OF IRISH RIVERS

James C.I. Dooge

Centre for Water Resources Research,
University College Dublin

INTRODUCTION

Rivers have played such a major role in human life down through the ages that they have entered deeply into human thought in a variety of ways. To speak of rivers inevitably involves a preliminary decision as to the manner in which the discourse is to be limited if its delivery is to be completed in a finite time. The preliminary limitation inherent in the invitation to me to speak in a seminar on Irish rivers and to deal with quantitative aspects of river flow does not resolve this problem because a wide variety of topics still remain. I have chosen to speak on only four: historical accounts of the flow of Irish rivers from the Leabhar Gabhála to the 1851 Census, the early attempts at flow measurement from Alexandria in the first century A.D. to Dijon in the seventeenth, the contribution of Irish engineers in the nineteenth century to modern methods for estimating river flow, and the present state of the art of flow measurement and estimation in the Ireland of today. It is hoped in this way that the various aspects of the biology of Irish rivers treated in the papers in this symposium can be placed in the broader context of Irish rivers as viewed by a number of the disciplines represented in the Royal Irish Academy.

HISTORICAL ACCOUNTS OF IRISH RIVER FLOW

Celts and the cult of water

In prehistoric times, the world-view of most peoples included the idea of a vast lake below the surface of the ground, which was both the passage to the afterlife and the source of knowledge derived from the gods of the Other-world. This vast underground lake was considered as the main source of springs and of the flow of rivers. Wells and rivers were a centre for ritual among the Celts in Ireland, in Britain and on the mainland of Europe.

Thus Anne Ross (1967, 46-7) in her *Pagan Celtic Britain* writes of this central importance of water for the Celts:

"Springs, wells and rivers are of first and enduring importance as a focal point of Celtic cult practice and ritual. Rivers are important in themselves, being associated in Celtic tradition with fertility and with deities such as the divine mothers and sacred bulls, concerned with this fundamental aspect of life....

The Celtic mother-goddesses, who frequently also function in the role of war-goddesses and prognosticators, have a widespread association with water. This is due, no doubt, to their own obvious connection with fertility which, in the popular mind, could be likened to the life-giving powers of water which could be witnessed by man himself."

Duval in his "Les Dieux de la Gaule" (1976, 60-1) writes to the same effect:

"Les eaux, richesse et agrément d'un pays, étaient heureusement et harmonieusement distribuées en Gaule sous toutes leurs formes et y recevaient des cultes divers. Des fleuves sont nommément divinisés: Rhin (*Rhenus pater* sur une inscription), Seine, Marne, Yonne et Saône. Les eaux courantes devoirent prendre dans l'imagination populaire la forme de démons, animaux, monstrueux ou humains et il semble qu'en Gaule et dans autres pays, la croyance prévalait que dans les rivières habitaient non des dieux mais des déesses."

In Britain, the names of such important rivers as the Clyde, the Dee and the Severn reflect their association with the goddesses Clota, Deva and Sabrina.

In Celtic mythology, wells and rivers were regarded as the prime source for the tapping of the reservoir of wisdom and knowledge stored in the great lake of the Otherworld located at the centre of the Earth. Swimming in this lake were the salmon of wisdom who grew fat and wise by eating the nuts that fell from the mystic hazel trees that overshadowed the great lake (O'Rahilly 1946, 322-3). Once in seven years either the nuts or the salmon escaped into the River Boyne and were a source of wisdom to any mortal who found them.

The origin of Irish rivers

There are a number of traditional references in early Irish literature to the formation of lakes and rivers. Leabhar Gabhála Éireann (The Book of Invasions) concerns itself with the origins of a number of physical features and the names attached to them. According to the edition of R.A.S. Macalister (elected a Member of this Academy in 1910 and elected President in 1926):

"Secht loch-thomadmand fo thír nÉrenn in-aimsir Parthaloin".

These seven lake bursts were a substantial addition to the three lakes and nine rivers said to have existed in Ireland in the period between the destruction of the first inhabitants in the biblical Flood and the settlement of the island by Parthalon and his group. The Leabhar Gabhála mentions four lake bursts in the time of Nemed over three hundred years later and three more in the time of the Tuatha Dé Danann (Macalister 1937). Most of these eruptions of underground water are associated in the literature with the digging of a grave for a king or a hero and consequent tapping of the great underground reservoir of the Otherworld.

The alternative view of precipitation (hail, rain, or snow) as the source of springs and rivers is also suggested in early Irish literature. One story makes Uisneach, which was a twin centre with Tara in Celtic Ireland, the source of the Irish river system. This account tells of how a great hailstorm fell on those assembled at Uisneach for the inauguration as king of Diarmaid, son of Carball:

> "Such was its greatness that the one shower left twelve chief
> streams in Ireland for ever".

This story was later Christianised under the form of a miracle by Saint Ciarán who broke a prolonged drought by causing twelve rivers to radiate from Uisneach.

The oldest documentary account of Irish geography is that due to Ptolemy who lists fifteen rivers. The names given and their most probable modern equivalents are: Logia (Lagan), Buvinda (Boyne), Oboka (Liffey), Modonnos (Avoca), Birgos (Barrow), Dabrona, probably a corruption of Sabrona (Lee), Iernos (Roughty near Kenmare), Dur, probably a corruption of Dub (Laune or Maine), Senos (Shannon), Librios (no trace of Irish name), Ravios (Erne or Roe in County Derry), Vivda (probably Bann), Argita (Bush in County Antrim). The above identifications (O'Rahilly 1946) are due to T.F. O'Rahilly, who was elected to this Academy in 1914.

The origin of the River Boyne

Owing to the complications of the Quaternary Ice Age, the geomorphological analysis of the origins of the major Irish rivers is not straightforward. The more poetical accounts in Celtic mythology are much more direct and uniform and correspond to similar accounts found in other ancient traditions. A typical account is that of the origin of the River Boyne as a result of the pride and the punishment of the goddess Boand, wife of Neachtain and later mother of Oengus through a union with the Dagda, one of the omnicompetent deities of the Tuatha Dé Danann.

The legend of Boand is beautifully presented in the Metrical Dindshenchas. The version quoted below is taken from the 1913 edition by Edward Gwynn (Member of the Academy in 1896 and President in 1934) published by the Academy in the Todd Lecture Series. The poem opens with a discussion of the many names given to the river:

> "Sid Neachtain is the name of the mountain here
> the grove of the full keen son of Labraid,
> from which flows the stainless river
> whose name is Boand ever-full.
>
> Fifteen names, certainty of disputes,
> given to this stream we enumerate,
> from Sid Neachtain away
> till it reaches the paradise of Adam."

Having given the local names of stretches of the river, the poem identifies with other rivers of the known world:

> "Severn she is called
> through the land of the sound Saxons,
> Tiber in the Romans' keep:
> River Jordan thereafter in the east
> and west River Euphrates.
>
> River Tigris in enduring paradise,
> long is she in the east, a time of wandering
> from paradise, back again hither
> to the streams of this Sid."

This picture of a hydrological cycle centred on a vast underground reservoir with rivers breaking through the ground surface and returning to the reservoir was dominant in ancient times. It is more likely that its adoption in Ireland reflects an earlier attempt to explain the river systems in the karstic limestone regions of central and southern Europe rather than observation of typical conditions in the post-glacial Irish landscape.

Having described the secret well which was in the charge of Neachtain, the poem describes how Neachtain's wife Boand defied the magic powers of the well:

> "Hither came on a day white Boand
> (her noble pride uplifted her),
> to the never-failing well
> to make trial of its power.
>
> As thrice she walked round
> about the well heedlessly
> three waves burst from it
> whence came the death of Boand".

The three waves struck a foot, an eye and a hand so that Boand in her hampered flight produced the meandering course of the river.

> "She rushed to the sea (it was better for her)
> to escape her blemish
> so that none might see her mutilation;
> on herself fell her reproach.
>
> Every way the woman went
> the cold white water followed
> from the Sid to the sea (not weak it was),
> so that thence it is called Boand."

This account of the origin of a meandering river is certainly more intellectually soothing than the modern theories of sediment transport, regime channels, graded rivers and entropy production in landscape evolution.

Historical floods and droughts in Ireland

The medieval annals contain numerous references to floods and droughts in Ireland. In the compilation by Britton (1937) entitled "A Meteorological Chronology to A.D. 1450", nearly all of the earliest entries refer to the Irish annals. Apart from the biblical Flood, the earliest reference to flooding is the overflowing of Loch Con and Loch Dechat in the twelfth year after the coming of Partholon to Ireland (about 2670 B.C. according to one chronology). More extensive flooding is ascribed to a later event about 1630 B.C. when nine lakes erupted. The report that about 800 B.C. there was abundant grain and fruit despite a lack of rain for ten years is hardly reliable.

A.P. Smyth (1972) in a paper to this Academy places the dividing line between borrowed material and local contemporary observations in the annals at the middle of the sixth century A.D. Even after this date, the chronologies of the various annals are occasionally offset by one or more years and are also subject to random errors in dating. Some general indications of climate variation, and hence of river flow, can be obtained even for the earlier centuries. The Irish annals for the second half of the sixth century record warm dry weather in A.D. 583 ("intense heat in this year") and 588 or 589 ("a scorching and dry summer"), but there is no reference to such weather at any time during the seventh century. A life of St Columba refers to "a very great, continuous and severe drought in the spring time" about 668 which was ended when the saint's relics were exposed. This drought may have been confined to Iona though serious prolonged drought is usually quite widespread.

The eighth century seems to have been a relatively dry period. No fewer than fifteen serious droughts were recorded by Irish and British annalists between 713 and 775. The following is a sixteenth-century translation (Murphy 1896) of the account in the now lost Annals of Clonmacnoise for 759 of the breaking of one of these droughts:

"There was a Great famine throughout all the kingdome in the
beginning of his raigne, In soe much that the King himself has
very little to live upon, and being then accompanied with seven
godly Bishops, fell upon their knees, where the King very piti-
fully before them all besaught God of his infinite Grace and
Mercy if his wrath otherwise could not be appeased, before he
saw the Destruction of so many thousands of his subjects and
friends that were helpless of a releefe and Reday to Perrish, to
take him to himself for maintenance of his service, which
request was noe sooner made than a Great shower of silver fell
from heaven, whereat the King Greatly Rejoyced and yett (said
he) This is not the thing that can Deliver us from this famine
and eminent Danger, with that he fell to his Prayers againe; and
then a second of heavenly honey fell, and then the king said
that Great thanksgiving as before, with that ye third shower fell
of pure wheat, which covered all the fields over that like was
never seen before, soe that there was such plenty and aboun-
dance of wheat, that it was thought yet it was able to maintaine
many kingdomes. Then the King and the seven Bishopes gave
great thanks to the Lord".

The same story appears under the year 763 in the Annals of Ulster in the
metrical form:

"Tri frossa Ard uilline,
ar ghradh De do rimh,
Fross airgitt, fross tuirinne,
agus fross do mil."

The translation by Hennessy (elected a Member of the Academy in 1865)
reads:

"Three showers at Ard-uilline
fed through God's love from heaven:
a shower of silver, a shower of wheat,
and a shower of honey".

Apart from the four-year discrepancy in dates, the metrical version is so suc-
cinct and allusive that we would have difficulty without the longer Annals of
Clonmacnoise version in relating the incident to the ending of a drought and
famine. In fact, the two accounts may not refer to an actual drought but
may both represent a poetical explanation of the name of Niall Frosach (Niall
of the Showers) who in the eighth century was king of Muredach
in Inishowen and later high-king.

In the next few centuries the annals record both floods and droughts
with an increasing amount of realistic detail. In the tenth century the
Annals of the Four Masters record the extent of flooding at Clonmacnoise on
a number of occasions, while droughts were recorded for five of the eight

years between 987 and 994. In the twelfth century, the Annals of Innisfallen record the summer of 1129 as so torrid that the streams of Ireland were dried up, while Benedict of Peterborough records the disruption by flooding of Henry the Second's campaign in Ireland in October 1171; later in 1178 the Annals of the Four Masters record that

> "the river of Galway was dried up for several days, so that all things lost in it from time immemorial were recovered".

In the fourteenth century, it was recorded of 1326 that

> "that was an unprecedented drought in Ireland... so that the brooks and great rivers (which always gave plenty of water) were almost dry".

More specific information is contained in a reference to flooding at Kilkenny later in the century.

> "Also on Tuesday, the 15th of the Kalends of December, there was a very great inundation of water, the like of which had not been seen for forty years previously in which bridges, mills and buildings were overthrown and carried away; the waters did not reach to cover the foot of the great altar or the steps of the altar of the whole abbey of the Friars Minor at Kilkenny".

Not only is this a picturesque account but it has hard information on date and some indication of the recurrence period and level of flooding.

For later centuries the annals cease to be available and are replaced as sources by diaries, letters, newspapers, and official documents. The following is a typical entry in a diary for 1726 (Dixon 1953):

> "By the great rains and melting of the snow on the mountains with the violent sleeting rain that fell this January the 1st the waters began to rise in St. Patrick Street about 6 o'clock in the evening and continued doing so until 4 the next morning when it (sic) began to abate and at 4 p.m. was quite gone. It was near 7 feet in the middle of the street but less in the houses, in my kitchen I had 4 foot of water... River Liffey over-spread Barrack Street, drove in the City stables and rose so high that 37 horses belonging to the city scavengers perished".

The remarkable Part V of the post-famine Census of 1851 on Tables of Deaths compiled by William Wilde (elected a Member of the Academy in 1839) contains a listing of "Excessive rains and floods" from A.D. 634 to 1851 and a listing of "Droughts and heats, hot summers and mild winters from 3294 A.M. to 1851 A.D." (Wilde 1856). These listings reveal such features as the alternation of droughts and floods for most of the eighteenth century:

"half of Limerick drowned in 1705;

great inundations all over Europe in 1726;

extremely hot summers in 1736, 1737, 1740 and 1741;

another great flood and inundation in Limerick in 1742;

dry summers and drought in 1760, 1761 and 1762 combined
with autumn and winter floods at Cork and Dublin in 1761 and
flooding and destruction of bridges in Dublin 1762;

great floods in many parts of the south in January 1765 followed by extreme drought in summer;

flooding in every year between 1773 and 1777 and again in
1782, 1784, 1786, 1787, 1788 and 1792;

in 1794 the driest summer since 1733 was followed by a wet
autumn and in November boats plied for hire in the Lower Castle Yard."

A number of the entries in Sir William Wilde's listing for the eighteenth and
early nineteenth centuries are quantitative rather than qualitative, a reflec-
tion of the fact that in climate records the anecdotal was being replaced by
the instrumental.

RIVER FLOW AS A QUANTITATIVE CONCEPT

River flow as volume per unit time

Between 3000 B.C. and 2000 B.C. there was an expansion and flowering
of a new type of civilisation in four widely separated centres: the Nile valley
in Egypt, the alluvial plain between the Euphrates and the Tigris, Marrapan
and Makenjo in the Indus valley, and in the valley of the Huang Ho. In all
cases the new development was founded on the control of the river flow
through irrigation, drainage and flood protection. These activities gave rise
to new legal concepts as recorded in the Code of Hammurabai (1700 B.C.)
and to new forms of administration. In the latter connection, Wittfogel
(1956; 1957) has argued that the requirements for strict water control in the
public interest inevitably gave rise to the first examples of bureaucratic government and of central planning in these hydraulic civilisations. All of the
major hydraulic engineering works in these civilisations were designed without the help of any adequate concept of how the flow of water could be quantified. Even more remarkably, later structures of great sophistication such
as the qanats for supplying water from subsurface sources in Armenia and
Persia (1000 B.C.) and the elevated aqueducts supplying water from springs
in Anatolia (800 B.C.) and Rome (300 B.C.) were efficiently designed without

formulating flow as a physical entity with the dimension of volume per unit time. Thus the design of the elaborate Roman aqueducts and the checking of their discharges to detect illegal abstraction were founded on the notion that a water discharge was represented solely by the cross-section of flow.

The concept of flow as volume per unit time is first found in the writings of Hero of Alexandria (*floruit* A.D. 60) who discussed in his *Dioptra* the problem:

> "Given a spring, to determine its flow that is
> the quantity of water which it delivers".

Hero recommends the diversion of the stream into a reservoir at a lower level and the measurement of the volume of inflow in a given time as measured by a sundial. He states specifically that the area of cross-section of the stream itself need not be measured. This is essentially the same as the method used today to calibrate flow-measuring devices in a hydraulic laboratory. The measurement of the flow in a large stream or river in this way is not practicable and only became so when the flow was recognised as the product of area and velocity. Hero was aware that the flow in any channel would increase with the velocity but it is not clear from his text whether he regarded the quantity of flow as the direct product of velocity and cross-sectional area. However, his grasp of the hydraulic principle of continuity (inflow minus outflow equals change in storage) was quite clear. Unfortunately his work was ignored by his contemporaries and by later generations.

The clear description of flow as velocity times area, as well as the recognition of the continuity of the amount of flow from section to section, appears in a number of places in the notebooks of Leonardo da Vinci (1452-1519). Leonardo's writings on hydraulics only slowly became available to scholars after the compilation by Arconati in 1643 of the writings on water contained in the twelve volumes of manuscripts deposited by him in the Ambrosian Library in Milan in 1637. Actually the more specific references to the principle of continuity are not contained in this compilation but are to be found in the Leicester Manuscript (now the Codex Hammer) which deals largely with water (Calvi 1923).

The key publication which brought the principle of continuity to general notice was the *Della misure dell'acque correnti* published in 1626 by Antonio Castelli (1578-1643), a student and collaborator of Galileo and hydraulic consultant to Pope Urban VIII. The third edition of 1660 was translated into English in 1661 and into French in 1664. The first part of the book is devoted to arguments based on analogy (taps emptying a barrel, cords drawn through a hole of equal bore, drawing or spinning of metal wire) and the second part is deductive, taking the form of six mathematical propositions. After the publication and translation of Castelli's work, sixteen centuries after that of Hero of Alexandria, the discharge of a stream was now firmly established as the quantity of volume per unit time given by the product of the cross-sectional area and the mean velocity of flow.

Even today the direct measurement of the flow in a stream is only carried out occasionally. This is done to calibrate at suitable locations along a river the relationship between water level and flow so that a continuous record of water level can be transformed into a continuous record of flow. In the absence of a record of water level, methods of varying sophistication are used to derive an estimate of flood flow from storm rainfall. Water levels and rainfall have been recorded for thousands of years but not for hydrological purposes. The Nilometer on Elephantine Island near Assuan erected four thousand years ago was fixed on the wall of a covered corridor accessible only to the priests and royal advisers since the level of the Nile was a major determinant of the gross national product and the basis of the level of taxation. The earliest available reference to rain gauges occurs in a book on politics and administration by Kautilya (400 B.C.) because the amount of the monsoon rainfall was the basis of the land tax in that part of India.

Theories on the origin of streamflow

Reference has already been made to the widespread belief in early times in springs arising from a vast underground lake which was the source of wisdom and of contact with the Otherworld. This view remained dominant through the rise of classical philosophy, medieval technology, Renaissance thought, and down to the eighteenth century. Thales (624-546 B.C.), designated by Aristotle as the first philosopher, held that the earth rested on water which was "the origin of all things". Thales and his school believed that elevated ground acted like a sponge in raising water above its base level and making it available to supply springs and rivers (Garbrecht 1987). Aristotle himself (384-322 B.C.) taught that the supply of water for streams was produced by the condensation of air to water in fissures in the earth. Pliny (A.D. 23-79) suggests that the subsurface water is pushed upward by the air in a type of siphonic action. A similar view is found in Leonardo (1452-1519) who writes of the circulation of water in the earth resembling the flow of blood in the human body and compares a spring on a mountain to blood flowing from a broken vein in the forehead. The view of underground water as the main source of springs and rivers persisted throughout the seventeenth century and to a lesser extent in the eighteenth century and was only displaced by the modern view of precipitation as the main source of streamflow by field measurements based on the concept of flow continuity enunciated by Castelli.

Isolated instances of support for the atmospheric origin of streamflow can be found among the many theories put forward down the centuries but none of them gained wide support. These instances include Xenophones of Colophon (570-475 B.C.) and Diogenes of Apollonia (460-390 B.C.) in classical times as well as Besson (1500-1590) and Palissy (1510-1590) at the end of the Renaissance. Palissy is better known as an innovative potter, a pioneer geologist, and a persecuted Huguenot. Palissy's book entitled *Discours admirables de la nature des eaux et des fontaines tant naturelles qu'artificielles* takes the popular Renaissance form of a dialogue between Theory and Practice. In the dialogue Practice maintains that springs and rivers originate from rainfall and not from seawater or from condensed air within the elevated land mass. He adduces arguments for his case such as the fact that periods of drought show a better connection with periods of low rainfall than with periods of high tide and suggests that the quantity of vapour available for condensation would be insufficient to supply the discharge of the world's rivers. The opposing theories of the origin of streamflow remained locked in a competition of assertion and counter-assertion for another two centuries after the appearance of Palissy's work. The argument was only resolved, and then only gradually, by the use of measurements to resolve the central question of whether precipitation was sufficient to account for the observed streamflow.

The first comparison of rainfall and streamflow

In 1674 a book, *L'Origine des fontaines*, by an anonymous author was printed in Paris. It contains a calculation intended to show that "The rain- and snow-waters are sufficient to make springs and rivers run perpetually." A lengthy abstract from this book appeared in the *Philosophical Transactions of the Royal Society* in November 1675. A second edition of the book appeared in 1678. The quantitative argument was improved on by Mariotte a short time later but no similar comparisons appear to have been made for at least 100 years.

The author of this 1674 work based his computation on a comparison between the flow of the upper Seine at Aignay-le-Duc some three leagues below its source and the local rainfall, presumably measured at Dijon. He gives the measured annual rainfall as 18 7/12 inches for 1668/69, $8^1/_2$ inches for 1670/71, and $27^1/_2$ inches for 1673/74. He then calculates (incorrectly) the average of these as 19 7/36 inches. This is only one of three instances of faulty arithmetic in the calculation. The nearest streams on either side were found to be 2 leagues away so he took the catchment area as 6 square leagues or 31.25 million square fathoms. The author then estimates the combination of an annual rainfall of 19 7/36 inches on 6 square leagues as giving an annual volume of rainfall of 224,899,942 muids. Since one inch on one square fathom equals 3 cubic feet and there are 8 cubic feet in a muid this total should have been 224,934,896 muids but the difference is not significant. The problem now resolved itself into the question of whether this amount was sufficient to account for the annual flow of the stream at Aignay-le-Duc.

To estimate the runoff from the upper Seine, the author calculated the flow to be 24 times that of the small Gobelins River near Versailles which had been measured as 50 inches. The author mentions that an inch of water (used as a measure of discharge) has been held as equivalent to anything from 70 muids to 144 muids (in 24 hours) and specified his own figure as 83 muids for 24 hours. This choice would give to the annual volume of flow of the Upper Seine the figure of 83 x 1200 inches x 365 days = 36,354,000 muids. With his usual penchant for small arithmetical errors, the author gives the figure as 36,453,600 muids.

A comparison of the above estimates of annual volume of rainfall and annual volume of runoff gives the runoff as only one-sixth of the rainfall and thus amply accounted for by it, even allowing for the uncertainty of the

Seine/Gobelins ratio of 24 and the inch-flow equivalent as 83 muids in 24 hours. The author concludes by saying:

> "I am well aware that this deduction is not sure but who can give a surer? However such as it is, I think it is more satisfactory than a bare negative as is that of those who pretend it rains not enough to furnish sufficient quantities of water for the constant running of rivers."

We can agree with the author in these sentiments.

But who was this anonymous author? There is probably sufficient evidence (Dooge 1959; Tixeront 1974) to attribute it to Pierre Perrault, one of four well-known brothers (Hallays 1926). Nicholas was a noted theologian prominent in the Jansenist disputes; Claude was a naturalist and author as well as architect of the Louvre; Charles played a large part in the foundation of the Academie Française and is also remembered for his writing, including fairy stories such as "Cinderella". Pierre Perrault (1608-1680) followed his father in training as a lawyer and was appointed Receiver General of Taxes for Paris under the administration of Colbert. An unexpected remission of taxes by Louis XIV to celebrate a military victory resulted in Perrault and other tax-farmers, who had already bid for the right to gather taxes in the following year, being made bankrupt. Perrault was allowed to retire to Dijon which was the scene of the rainfall-runoff comparison described above. His interest in the origin of springs may have been prompted by the discussion in Chapter 3 of Book VIII of Vitruvius (written 25 B.C.) whose *Ten Books on Architecture* were translated into French by his brother Claude Perrault and published in 1673. The author of the 1674 book shows no sign of being acquainted with Castelli's work, the French translation of which had appeared in 1664.

Streamflow estimates based on area and velocity

The first crude estimation of the rainfall-runoff ratio by Pierre Perrault (1674) was quickly followed by the much more scientific comparisons by Mariotte (1686) and Halley (1687; 1691) and then over a century elapsed before the next significant paper, that by Dalton in 1802. Mariotte's improved version of Perrault's 1674 comparison of rainfall and runoff must have been made within a short period since Mariotte died in 1684. An account of this more scientific comparison, which explicitly uses the concept of streamflow as a product of area and velocity, is a small section in Mariotte's landmark book *Traite du mouvement des eaux et des autres corps fluides*. The first major improvement relates to the size of catchment area which Mariotte takes as the Seine above Paris estimated as 60 leagues in length and 50 leagues in breadth. From personal measurements at Dijon, he estimated the annual rainfall as 17 inches and for calculation purposes reduced this to the "safe" figure of 15 inches. The combination of 15 inches of rainfall on an area of 3000 square leagues gives an annual volume of rainfall of 714,150 million cubic feet. To estimate the runoff, Mariotte observed the Seine above the Pont Royal and took the width as 400 feet and depth (which varied from 2 feet to 10 feet) at an average depth of 5 feet. He observed that in flood time a stick carried in mid-stream moved at the pace of a very fast walker which he assumed to be about 250 feet per minute. For the average conditions of a depth of 5 feet, Mariotte took the surface velocity as 150 feet per minute and the average velocity over the cross-section as 100 feet per minute. This gave him 400 x 5 x 100 = 200,000 cubic feet per minute as the flow at average conditions. This corresponds to an annual volume of flow of 105,120 million cubic feet, which is less than one-sixth of the estimated annual rainfall. Thus Mariotte's more careful comparison gave much the same general result as that of Perrault.

Edmond Halley, better known as an astronomer and as the editor of Newton's *Principia*, published a number of papers on hydrologic topics (Dooge 1974). In a paper in the *Philosophical Transactions of the Royal Society* for September/October 1687, Halley describes some experiments on evaporation conducted by him at Gresham College and used it as a basis for estimating the water balance of the Mediterranean. He measured the evaporation as one-fiftythird of an inch in two hours, which he extrapolated to one-tenth of an inch over a 12-hour day; applying this to the area of the Mediterranean taken as 160 square degrees, he computed the daily volume of evaporation as 5280 million tons of water. Halley then estimated the flow of the Thames at Kingston Bridge as 20.3 million tons per day on the basis of a cross-sectional area of 300 square yards and a velocity of 2 miles per hour. As a next step, he approximated the inflow into the Mediterranean from large and small rivers as equivalent to nine large rivers each having ten times the flow of the Thames. The resulting total runoff is thus estimated as 1827 million tons per day, which is only one-third of the estimated evaporation.

Halley ends this 1687 paper by stating that further discussion "is intended, with leave, for a farther entertainment of this honourable company". In fact the sequel only appeared nearly four years later in *Philosophical Transactions* No. 192, which was prefaced by the statement:

> "The publication of these transactions having been
> for some time past been suspended, chiefly by rea-
> son that the unsettled Posture of Public Affairs did
> divert the Thoughts of the Curious of Matters of
> more immediate concern than are *Physical* and
> *Mathematical Enquiries*."

These matters that diverted the thoughts of the curious included the Whig Revolution of 1688 and the military defeat of James II at the river created by the rushing goddess Boand in her flight. In his second paper Halley rejected Mariotte's conclusion and returned to Aristotle's theory of the condensation of dew inside mountains.

John Dalton (1766-1844), though remembered largely for his fundamental contributions to chemistry, maintained a life-long interest in meteorology. In the first paragraph of a paper read to the Literary and Philosophical

Society of Manchester in 1799 and published by them in 1802 he writes:

> "Naturalists, however, are not unanimous in their
> opinion whether the rain that falls is sufficient to
> demands of springs and rivers, and to afford the
> earth besides such a large portion for evaporation
> as it is well known is raised daily."

This is a clear indication that the question of the origin of the flow in springs or rivers remained an open question more than two centuries after the persuasive arguments of Bernard Palissy and more than a century after the plausible quantitative comparisons of Perrault and Mariotte.

The paper by Dalton proceeded to examine the question by calculating an annual water balance for England and Wales. On the basis of thirty rain-gauges, he estimated the mean annual rainfall for England and Wales as 31 inches. To this he added 5 inches per annum for precipitation in the form of dew but in his final balance assumed all this to be evaporated. Dalton based his estimate of river flow on a revised estimate of the flow of the Thames used by Halley in his paper of 1687. Dividing the remainder of England into six districts, he calculated the area of each and took its shape and the nature of the river system into account in order to estimate the ratio of the runoff from each district to that of the Thames. In this way, he computed the discharge of all the rivers of England and Wales as nine times that of the Thames and equivalent to a depth of 13 inches.

Dalton then attempted to close the water balance by an estimate of 25 inches as average annual evaporation based on three years' measurement on a 10-inch diameter evaporimeter. Dalton summarised his general conclusion as:

> "Upon the whole then I think that we can finally
> conclude that the rain and dew of this country are
> equivalent to the quantity of water carried off by
> evaporation and by the rivers. And as nature acts
> upon general laws, we ought to infer, that it must
> be the case in every other country *until* the con-
> trary is proved."

Dalton attributed most of the lack of balance (31 - 13 - 25 = - 7) between the components to an overestimation of evaporation and this view is supported by a review of his attempted water balance in the light of modern measurements (Rodda 1963).

CONTRIBUTIONS OF IRISH ENGINEERS

Hydraulic engineering in Ireland

The past two centuries have seen a good deal of activity in Ireland in relation to major works of hydraulic engineering. For many centuries before that, small engineering structures such as stepping stones and causeways, fishing weirs on both non-tidal and tidal rivers, and waterwheels of increasing complexity had been designed and constructed according to local tradition and practice. During the eighteenth century the major innovations were in inland navigation, during the nineteenth century in drainage and flood control, and during the twentieth century in hydroelectric development and water resources planning.

Some of the works carried out under a 1715 Act of the Irish parliament attracted attention outside this country. The purpose of this Act was to encourage "the drainage and improvement of bogs and of profitable low grounds" and to facilitate "the despatching of inland carriage and conveyance of goods from one part to another within the Kingdom". It was under this Act that the Newry Canal was commenced with the objective of conveying coal from the area around Lough Neagh to Newry and thence by sea to Dublin. This was the first summit level canal in these islands and as such of considerable historical importance. Frisi in his important work entitled *Del mode di rigolare i fiumi e i torrenti* published in 1762 includes a 45-page "Essay on Navigable Canals" which includes the following reference: "In Ireland, they are constantly extending the navigation of the river Shannon into the heart of the Kingdom". Extensive works under the 1715 Act were only carried out in the latter half of the eighteenth century when surplus funds became available in the Irish exchequer. Canal works were encouraged by Charles Vallancey, who in 1763, shortly after his appointment as engineer in ordinary for Ireland, published a *Treatise on Inland Navigation.* Vallency was a founder Member of this Academy and notorious as a controversialist on such subjects as the origin of the Celts.

In the nineteenth century, the schemes of navigation and drainage promoted by local undertakers gave way to schemes of drainage and navigation planned and executed as public works, with such features as power of entry to lands for survey and works, compensation for weirs and mills affected at a level ultimately determined by central government, and a quasi-permanent corps of engineers. A full history of the decade of operation of the 1842 Arterial Drainage Act would throw a good deal of light on many important aspects of both engineering development and the conflict of social ideas in mid-nineteenth-century Ireland. A short account is given in the next section of some contributions by the Irish engineers to the estimation of river flow which remain valid for engineering practice today.

In the first half of the twentieth century, quantitative information on the flow of certain Irish rivers was required for the efficient planning of hydroelectric development in addition to the information required for the design of arterial drainage schemes under the 1925 Act and later the 1945 Act. In the second half of the century, the increasing demand for water (domestic, agricultural and industrial) created a further demand for information on river flow for the planning of regional water resources and the design of individual water supply schemes. The final section of this paper outlines the development of these hydrometric surveys which provide the basis of our present knowledge of the quantitative aspects of the flow of Irish rivers.

The nineteenth-century engineers

The contribution of the Irish engineers to Irish development and to hydraulic engineering in the nineteenth century can be appreciated by considering the careers of three of them who are still remembered today: William Mulvany (1806-1886), Thomas Mulvany (1821-1892) and Robert Manning (1816-1897). They represent the first generation of Irish engineers to rival in prominence English and Scottish engineers working in Ireland.

William Mulvany was the eldest son of Thomas James Mulvany, the biographer of Gandon and a founder and first curator of the Royal Hibernian Academy. William worked on the Ordnance Survey from 1825 to 1827, then under Richard Griffith on the Boundary Survey from 1827 to 1835 when he was transferred to the employment of the Shannon Commissioners at the request of Colonel John Fox Burgoyne who was Chairman of that body. John Fox Burgoyne was the natural son of General John Burgoyne, whose defeat at Saratoga Spring had the important effect of influencing France, Spain and Holland to ally themselves with the American colonists. John Fox Burgoyne was the first Chairman of the Commissioners of Shannon Navigation, the first Chairman of the Irish Board of Works, the first President of the Institution of Civil Engineers of Ireland, and as Field Marshal Sir John Fox Burgoyne was Inspector General of the Royal Engineers at the time of the Crimean War. He was elected a Member of this Academy in 1834. Burgoyne with the assistance of William Mulvany drafted the 1842 Arterial Drainage Act and lobbied parliament to ensure its passage. William Mulvany at the age of 36 was appointed as Commissioner for Drainage at a salary of £400 per annum and as Commissioner for Fisheries at an additional salary of £200 per annum.

William Mulvany (who had been elected to the Academy in 1841) recruited a group of six or seven engineers to implement the radical provisions of the 1842 Act. The group had varying backgrounds and experience: John MacMahon, a public works contractor and a veteran of the canals era who retained his physical and intellectual vigour despite his 70-odd years; at least two other engineers with experience of the large-scale and varied works under the Shannon Commission; and Samuel Roberts, a 21-year-old graduate of Cambridge whose mathematical education represented a new approach to the training of an engineer. Between 1842 and 1846 surveys were completed and works commenced in 6 districts; in 1846 under the impact of the Great Famine the procedures for surveying a district were simplified and the required assent from landed proprietors reduced from two-thirds to one-half; between 1846 and 1851 over 400 districts were surveyed and works were commenced in 100 districts. After the immediate crisis had passed, the landowners protested against the drainage charges levied and a House of Lords Committee investigated the operation of the arterial drainage works and found in favour of the landowners. The Chairman of this House of Lords Committee was the third Earl of Rosse, the builder of the great telescope at Birr. William Mulvany was retired as Commissioner for Drainage on a pension of £1,000 a year. His loss to the cause of Irish development is indicated by the fact that he is remembered even to the present day in Germany as one of the key figures in the industrial development of Ruhr where he worked from 1855 until his death in 1886. There is a full-scale biography of him in German (Bloemers 1922).

In the absence of reliable flow records, the hydrologist often attempts to relate flood runoff to recorded storm rainfall. The simplest method of doing this, which is termed the rational method, is still used today for relatively unimportant areas of small catchments and is the origin of the more sophisticated methods used for important works on large catchments. The first formulation of the rational method is given in a paper read to the Institution of Civil Engineers of Ireland in February 1851 by Thomas Mulvany, the youngest brother of William Mulvany (Dooge 1987). This paper defines clearly and describes the importance of the concept of the time of concentration of a catchment that can still be used verbatim in lecturing on the rational method almost a hundred and forty years later. Thomas Mulvany had worked under his elder brother on the works on the Lower Shannon and transferred to the Office of Public Works as a drainage engineer during the famine period. He resigned from the latter position, forgoing his pension, to join his brother in the Ruhr in 1855. In 1878 he emigrated from Dusseldorf to New Zealand and in 1880 published a series of essays on the industrial development of that country.

The most widely used formula for calculating the flow at a given depth in a channel of given cross-sectional shape, bed slope and surface roughness is known throughout the world as the Manning formula. Robert Manning, having worked on the management of his uncle's estate in Waterford, was employed in January 1846 as a drainage clerk by Samuel Roberts who was also from Waterford. In April 1846 Manning was appointed accountant and draughtsman and in October 1846 assistant engineer. From 1848 to 1855 he acted as a district engineer. In 1847 he was engaged on surveys in the west of Ireland and must have seen some of the worst effects of the famine. In a report to the Board of Works, a district engineer reported that during meal breaks men who had no food with them would go behind a hedge so that their fellow workers would not be aware of this fact. From 1855 to 1866 he worked for the Marquess of Downshire on the survey of his Irish estates, the construction of Dundrum Harbour, and a water supply for the City of Belfast. He returned to the Office of Public Works in 1869 as Assistant Chief Engineer and was Chief Engineer from 1874 till his retirement in 1891 at the age of seventy-five. Manning is chiefly remembered today because of his paper, read in December 1889 to the Institution of Civil Engineers of Ireland, "On the flow of water in channels and pipes". Based largely on the classical series of experiments by Bazin (1865), which have not been replaced or superseded even by modern work, Manning put forward a monomial-type formula as the best fit to some 160 experiments. Even though similar formulae were proposed by Gauckler in 1867 and Strickler in 1921, this monomial formula, which indicates the velocity as proportional to the square root of the slope and to the two-thirds power of the hydraulic radius, is almost universally known as the Manning formula.

The work of the two Mulvanys and of Manning is still part of hydraulic engineering practice. In May 1989, the University of Virginia organised an

International Symposium on Channel Flow and Catchment Runoff. These two themes were chosen because 1989 was both the centenary year of Mannings's paper on open channel flow and of the first American paper on the rational method introduced by Thomas Mulvany. In December 1989 the Institution of Engineers of Ireland celebrated the anniversary of Manning's paper with a special symposium.

The modern hydrometric record

Even in the hectic conditions of the huge expansion of drainage works during the famine period, William Mulvany encouraged his engineering staff to use measured flood flows as the basis of design and as a check on their design methods (Dooge 1987). Remarkably high flood flows occurred in 1851 and 1852 when engineers were resident in over 100 drainage districts. The flows of the smaller streams were gauged by weirs and those of the larger streams by float measurements or by the use of a slope-velocity formula. The results of these measurements were assembled and tabulated at the head office of the Board of Works in the Custom House but unfortunately these tabulations have not survived. The existence of these flow measurements helped to refine the design methods used between 1843 and 1850 and to stimulate the work of Thomas Mulvany in measuring flows from small steep catchments and formulating the concept of time of concentration (Dooge 1987). Even in the lean years for the Board of Works and for Arterial Drainage in the latter half of the nineteenth century, some flow records were kept, as is evident from testimony by Robert Manning to a number of Royal Commissions.

Continuous records of water levels on the Shannon at Killaloe have been kept since 1893 (Chaloner Smith 1919; 1923) and were invaluable in the planning of the Shannon scheme. Modern hydrometric survey records in Ireland may be dated from the introduction of current meter measurements. Though the current meter was developed by Woltmann about 1790, it was not used in these islands until well over a century later. A pioneer in this regard was Michael A. Hogan (elected a Member of this Academy in 1939) who proposed the need for a hydrometric survey in Ireland in 1923 in a paper to the Institution of Civil Engineers of Ireland and introduced systematic techniques for current meter measurements in Britain (Hogan 1922; 1923; 1925; Dixon et al. 1937). J.A. O'Riordan of the Electricity Supply Board in papers to the Institution of Civil Engineers of Ireland in 1937 and 1949 presented the results of systematic measurements on the Shannon, the Liffey and the Erne.

In 1939, the Commission on Drainage recommended the establishment of a systematic and comprehensive hydrometric survey. In the 1940s the Office of Public Works erected hundreds of staff gauges and from the early 1950s onwards these were gradually replaced by autographic water level records. The Electricity Supply Board has continued to record water levels of rivers of interest to it. An Foras Forbartha has maintained a hydrometric network for the purpose of water resource planning since 1972. In Northern Ireland the responsibility for surface water measurements is divided between the Department of Agriculture and the Department of the Environment.

By agreement between the various agencies concerned, the island of Ireland has been divided into 40 hydrometric areas and the records for over 500 autographic water level recorders are maintained by the Water Resources Division of An Foras Forbartha and the Water Data Unit of the N.I. Department of the Environment. A *Surface Water Year Book* giving hydrographs of daily mean flow and flow durations has been published for the years 1975, 1976 and 1977 (A.F.F. 1978; 1981; 1982). In more recent years, An Foras Forbartha has published statistical analyses of river flows in the South-Eastern Region, i.e. hydrometric areas 11 to 17, and in the Southern Region, i.e. hydrometric areas 18 to 22 (A.F.F. 1984a; 1987). The Office of Public Works has also published summaries of its processed records giving monthly mean flows, highest and lowest flows (daily, monthly and annual), flow duration curves, and sustained low flows in wet, average, and dry years (O.P.W. 1982). The situation in relation to the measurement of river flow and other elements of the hydrological cycle in Ireland is summarised in a publication by the Irish National Committee for the International Hydrological Programme (IHP 1982).

The Office of Public Works has in the third quarter of this century sustained the nineteenth-century O.P.W. tradition of combining good engineering practice with innovative procedures. J.J. O'Kelly introduced the unit hydrograph approach, now recognised as a development of Mulvany's rational method, at a time when it was unknown in Europe (O'Kelly 1955). More recently the O.P.W. participated in the extensive research that led to the four-volume Flood Studies Report of the U.K. National Environmental Research Council. The application of the results of that study to Ireland was described in two papers to the Institution of Engineers of Ireland, one on floods (Cunnane and Lynn 1975) and one on low flows (Martin and Cunnane 1976).

The contribution of Irish hydraulic engineers has indeed been a worthy one. It is not for me to evaluate whether my own generation of river engineers in Ireland has lived up to tradition and has adequately trained the next generation in this field. What I can state is that the example of both the great drainage engineers of the nineteenth century and our immediate predecessors gave us good guidelines to follow.

REFERENCES

A.F.F. 1978 *Surface Water Year Book, Ireland 1975*. Dublin. Water Resources Division, An Foras Forbartha.

A.F.F. 1981 *Surface Water Year Book, Ireland 1976*, Volumes I and II. Dublin. Water Resources Division, An Foras Forbartha.

A.F.F. 1982 *Surface Water Year Book, Ireland 1977*. Hydrographs of Daily Mean Flows and Flow Duration Curves. Dublin. Water Resources Division, An Foras Forbartha.

A.F.F. 1984a *A Statistical Analysis of River Flows. The South-Eastern Water Resource Region*. Dublin. Water Resources Division, An Foras Forbartha.

A.F.F. 1984b *1984 Drought River Flows. A Comparison with Other Years.* Dublin. Water Resources Division, An Foras Forbartha.

A.F.F. 1987 *A Statistical Analysis of River Flows. The Southern Water Resource Region.* Dublin. Water Resources Division, An Foras Forbartha.

ARCONATI, L.P. 1643 *II Trattato del moto e Misura dell Acqua.* Codex Vaticanus - Barberini 4332.

BAZIN, H.E. 1865 Recherches expérimentales sur l'écoulement de l'eau dans canaux découverts. *Acad. Sc. Mem. sav. etr.* 19, 1-501.

BLOEMERS, K. 1922 *William Thomas Mulvany (1806-1885).* Veröffentlichungen des Archives für Rheinisch-Westfallische Wirtschaftsgeschichte Band VII. Essen. Baeleker.

BRITTON, C.E. 1937 *A meteorological chronology to A.D. 1450.* Geophysical Memoirs No. 70 (Vol. 8, No. 1). U.K. Met. Office Publications MO409A. HMSO.

CALVI, G. 1923 *I manoscritti di Leonardo de Vinci dal punto di vista cronologico storico e biografico.* Bologna.

CASTELLI, B. 1628 *Della misura dell'acqua correnti.* Rome. Nella Stamperia Canerale.

CHALONER SMITH, J. 1919 Notes upon the average volume of flow from large catchment areas in Ireland. *Trans. Inst. Civ. Eng. Ire.* 45, 41-118.

CHALONER SMITH, J. 1923 Notes on the rational method of eliminating flood effects from a series of river gorgings. *Trans. Inst. Civ. Eng. Ire.* 49, 126-59.

CUNNANE, C. and LYNN, M.A. 1975 Flood estimation following the Flood Studies Report. *Transactions Institution of Engineers of Ireland* Dublin 100 (February 1975), 29-44.

DALTON, J. 1802 Experiments and observations to determine whether the quantity of rain and dew is equal to the quantity of water carried off by the rivers and raised by evaporation, with an enquiry into the origin of springs. *Mem. Proc. Lit. Phil. Soc. Manchester* 5, part 2, 346-72.

DIXON, F.E. 1953 Weather in Old Dublin. *Dublin Historical Record* 13, 94-107.

DIXON, S.M., FITZGIBBON, G. and HOGAN, M.A. 1937 The flow of the River Severn 1921 - 36. *Journal of Institution of Civil Engineers* (London) 6, No. 7, 81-113.

DOOGE, J.C.I. 1959 Un bilan hydrologique au XVIIᵉ Siècle. *La Houille Blanche* (November 1959), 799-807.

DOOGE, J.C.I. 1974 The development of hydrological concepts in Britain and Ireland between 674 and 1874. *Hydrological Sciences Bulletin* XIX, No. 3, 279-302.

DOOGE, J.C.I. 1987 Manning and Mulvany. River improvement in 19th century Ireland. In G. Garbrecht (ed.), *Hydraulics and Hydraulic Research. A Historical Review,* 173-83. Rotterdam. A.A. Balkema.

DUVAL, P.M. 1976 *Les Dieux de la Gaule.* Paris.

FRISI, P. 1762 *Del modo di regolare i fiumi e i torrenti.* Lucca.

GARBRECHT, G. 1987 Hydrologic and hydraulic concepts in antiquity. In G. Garbrecht (ed.), *Hydraulics and Hydraulic Research. A Historical Review*, 1-22. Rotterdam. A.A. Balkema.

GWYNN, E. 1913 The Metrical Dindshenchas Part III. *Todd Lecture Series* 10, 26-9. Royal Irish Academy.

HALLAYS, A. 1926 *Les Perrault.* Paris.

HALLEY, E. 1687 An estimate of the quantity of Vapour raised out of the sea by the sun. *Phil. Trans. Roy. Soc.* 16, No. 189 (September-October 1687), 366-70.

HALLEY, E. 1691 On the circulation of the vapours of the sea and the origin of Springs. *Phil. Trans. Roy. Soc.* 17, No. 192 (January-February 1691), 468-73.

HENNESSY, W.M. 1887 *Annals of Ulster*, 230-1. Dublin.

HERO OF ALEXANDRIA 70 *Dioptra.* Vol. III *of Heronis Alexandrini Opera and supersunt omnia.* Leipzig. Five volumes, 1899-1914.

HOGAN, M.A. 1922 *Current meters for use in river gauging.* London. Department of Scientific and Industrial Research.

HOGAN, M.A. 1923 The development of Irish water resources and the need for a hydrometric survey. *Trans. Inst. Civ. Eng. Ire.* 49, 86-125.

HOGAN, M.A. 1925 *River Gauging.* London. Department of Scientific and Industrial Research.

HOUSE OF LORDS 1852 Report of Select Committee on the Operation of the Acts relating to Drainage of Lands in Ireland with Minutes of Evidence and Appendix. *House of Lords. Parliamentary Papers* 1852, XXI: 1.

I.H.P. 1982 *Hydrology in Ireland.* Irish National Committee for the I.H.P.

MACALISTER, R.A.S. 1937 *Lebar Gabála Érenn.* Dublin. Irish Texts Society.

MANNING, R. 1889 On the flow of water in open channels and pipes. *Trans. Inst. Civ. Eng. Ire.* 20, 161-6.

MARIOTTE, E. 1686 *Traité du Mouvement des Eaux et des autres Corps Fluides.* Paris.

MARTIN, J. and CUNNANE, C. 1976 Analysis and prediction of low flows and drought volumes for selected Irish rivers. *Trans. Inst. Eng. Ireland* 101 (March 1976), 21-8.

MULVANY, T.J. 1851 On the use of self-registering rain and flood gauges in making observations of the relations of rainfall and of flood discharges in a catchment. *Trans. Inst. Civ. Eng. Ire.* 4, Part II, 18-33.

MURPHY, D. (ed.) 1896 *The Annals of Clonmacnoise.* Royal Society Antiq. Ireland.

O'KELLY, J.J. 1955 Employment of unit hydrographs to determine the flow of Irish arterial drainage channels. *Proc. Inst. Civ. Eng.* London 4 (3), 365-412.

O.P.W. 1982 *Record of Flows: Summary of Processed Records.* Dublin. Hydrometric Section, Office of Public Works.

O'RAHILLY, T.F. 1946 *Early Irish History and Mythology.* Dublin Institute for Advanced Studies.

O'RIORDAN, J.A. 1937 The rainfall and runoff of the River Shannon at Killaloe from 1893 to 1936. *Trans. Inst. Civ. Eng. Ire.* 63, 149-85

O'RIORDAN, J.A. 1949 Runoff at the River Liffey and the Lower River Erne. *Trans. Inst. Civ. Eng. Ire.* 75, 87-123.

PALISSY, B. 1580 *Discours admirable de la nature des eaux et des fontaines tant naturelles qu'artificielles.* Paris.

PERRAULT, P. 1674 *L'Origine des Fontaines. Pierre le Petit.* Paris. (Translated with commentary by A. Larocque, 1967. New York. Hafner.)

RODDA, J. 1963 18th Century evaporation experiments. *Weather* 18, 264-9.

ROSS, A. 1967 *Pagan Celtic Britain. Studies in Iconography and Tradition.* London. Routledge and Kegan Paul.

SMYTH, A.P. 1972 The earliest Irish annals: their first contemporary entries and the earliest centres of recording. *Proc. R. Ir. Acad.* 72C, 1-48.

TIXERONT, T.J. 1974 L'hydrologie en France au dix-septième siècle. In *Three Centuries of Scientific Hydrology,* 24-39. Paris. UNESCO-WMO-IAHS.

VALLENCY, C. 1763 *Treatise on Inland Navigation.* Dublin.

WILDE, W. 1856 Report on Tables of Deaths. *Census of Ireland for 1851,* Part V, Vol. I, 342-6. Dublin. HMSO.

WITTFOGEL, K.A. 1956 The hydraulic civilisations. In W.L. Thomas (ed.), *Man's Role in Changing the Face of the Earth,* 152-64. University of Chicago Press.

WITTFOGEL, K.A. 1957 *Oriental Despotism: A Comparative Study of Total Power.* Yale University Press.

Part 2

Fish biology

In: Steer, M.W. (ed.) 1991 *Irish Rivers : Biology and Management*, pp 29-46. Royal Irish Academy, Dublin.

SALMON RESEARCH ON THE RIVER BUSH

Walter W. Crozier and Gersham J.A. Kennedy

Department of Agriculture for Northern Ireland, Fishery Research Laboratory, Coleraine, Co. Londonderry

ABSTRACT

The biology and management of Atlantic salmon (*Salmo salar* L.) have been studied on the River Bush since 1973, when the Department of Agriculture for Northern Ireland acquired the lease to the river and established hatchery and trapping facilities at Bushmills. Original objectives centred on salmon ranching experiments, but have been widened to encompass several topics:

1. investigations into stocking and juvenile salmonid ecology in nursery habitat;

2. evaluation of the techniques for the refurbishment of salmonid habitat following drainage schemes;

3. investigation of the freshwater stock/recruitment relationship of Atlantic salmon;

4. investigation of salmon stock enhancement on the R. Bush;

5. investigation of factors affecting the marine survival of Atlantic salmon.

Results from the early phase of the Bush research are summarised, together with recent data from longer-term studies on stock/recruitment and marine survival. The role of the R. Bush as an index river is described in the context of the future management of Irish salmon stocks.

D

INTRODUCTION

The R. Bush rises in Co. Antrim and is around 55 km in length, entering the sea on the north coast of N. Ireland. The total catchment is around 340 km^{-2} and the mean river discharge just over 5 cumecs (range, <1-90 cumecs).

Work on the salmon stocks of the river began in 1972, and was initiated by the Department of Agriculture for N. Ireland, who acquired a 30-year lease to the R. Bush and developed hatchery and trapping facilities at Bushmills. Angling on the river also devolved to the Department under the lease (with the exception of the lower 3 km, retained by the owner), and three stretches around Bushmills are managed on a day-ticket basis, with anglers required to submit returns, including scale samples and tag returns. In addition, three coastal netting sites were acquired under the lease, only one of which is operated regularly.

The original objectives of the Bush project centred on providing data for the management of a salmon river and on testing the economics of smolt rearing (Anon. 1973). To this end, facilities for trapping and counting wild smolts and returning adults were constructed at the Bushmills site using existing mill weirs and laydes. Adult fish ascending the river are deflected, via an electric barrier at the downstream weir, into a trap measuring 3 m x 3 m, with a nominal water depth of 1 m. This trap has a lifting floor to facilitate examination and removal of the trapped fish. Migrating smolts moving downstream are diverted into the site at an upstream weir and intercepted by a Wolff trap, leading directly into a holding tank with a lifting floor. Periodically, smolts are flushed into an illuminated observation table in which they are photographed for counting, thus avoiding excessive handling. An additional trapping facility at the upstream weir samples the overspill of smolts in spate conditions, when the weir can be overtopped.

Broodstock are held for stripping in separate enclosures accommodating up to 800 fish. The hatchery building houses 15 Heath Techna incubator units and 20 hatching/first feeding troughs. Twenty fibreglass early rearing tanks are sited adjacent to the hatchery. The river water supply to these tanks and the hatchery is part-filtered by a series of filter beds.

The main rearing area consists of 40 circular concrete tanks and a further 20 fibreglass early rearing tanks supplied with unfiltered river water. Although design capacity of 100,000 smolts has not been achieved (20-40,000 is more realistic) the extra capacity has been utilised in holding separate experimental groups of salmon.

The Bush project has evolved as a result of research carried out in relation to the original objectives. During the early years, the wild salmon runs were intentionally reduced to achieve low points on the stock/recruitment relationship, to provide broodstock for smolt rearing purposes in line with the ranching investigations and to meet financial constraints by commercial sale of fish. Results from this period indicate that the then stock level could not provide sufficient buffering against natural environmental variation. At that time ranching was not economically viable owing to poor survival of all farm-reared stages (Kennedy and Johnston 1986). The objectives of the project were reassessed in 1983 and 5 interrelated projects were established:

1. investigations into stocking and juvenile salmonid ecology in nursery habitat;

2. evaluation of the techniques for the refurbishment of salmonid habitat following drainage schemes;

3. an investigation of the freshwater stock/recruitment relationship of Atlantic salmon;

4. an investigation of salmon stock enhancement on the R. Bush;

5. investigation of factors affecting the marine survival of Atlantic salmon.

More recently, research has been initiated into establishment of a genetically marked stock of salmon at Bushmills for use in investigations into the genetic effects of salmon stocking. Comparisons of genetic variability in Bush wild and hatchery stocks have also been carried out (Crozier and Moffett 1989).

The adult and smolt traps have been in operation since 1973, providing quantitative information on the marine and freshwater survival of Atlantic salmon. Accordingly, the Bush is recognised as an index river by the International Council for the Exploration of the Sea (ICES). They have designated this and several other rivers throughout the north Atlantic to provide long-term scientific information on stock levels and exploitation rates for the database established by the ICES N. Atlantic Salmon Working Group. The Bush is now an integral part of an international research programme aimed at providing information on biology and exploitation of stocks of salmon, in order that management can be based on data from national contributions to mixed stock interceptory fisheries on the high seas and in home waters. The eventual aim is that this type of assessment will set permitted harvests in mixed and discrete stock fisheries at such a level as to ensure return to the rivers of sufficient numbers to meet target spawning levels.

RESULTS AND DISCUSSION

Salmon stocking investigations in nursery habitat

Investigations into the ecology of stocked juvenile salmon and resident brown trout (*Salmo trutta* L.) have been carried out in two small streams above Altnahinch dam in the head-waters of the R. Bush (see Kennedy 1981a and b; 1984a; 1985; Kennedy and Strange 1980; 1981a and b; 1986a and b; Kennedy and Johnston 1986). The work has covered a variety of topics, including the comparative survival of salmon from stocking with eyed ova, green eggs and swim-up fry and the extent of dispersal of stocked salmon fry. Investigations into the distribution of stocked salmon in relation to depth and gradient in nursery streams have been carried out, together with studies on the effect of intra- and inter-specific competition on growth, survival and distribution of stocked salmon and resident trout.

Investigations are presently in progress to determine the optimum stocking density of salmon fry in good nursery habitat. The results of this present study suggest that the smolt production from nursery habitat in N. Ireland, at average salmonid holding capacity, is in the order of 30 smolts 100 m^{-2} per annum. This is considerably higher than previously outlined in the literature, as based on natural ova deposition rates over whole river systems - where areas of unsuitable nursery habitat have been included (Kennedy 1988).

The findings from all these juvenile salmonid studies at Altnahinch have been applied in evaluating the production from nursery streams in terms of an economic model of salmon rivers in N. Ireland (Kennedy 1985; 1986; Kennedy in prep.).

The rehabilitation of salmonid habitat following drainage schemes

A pre-enhancement electrofishing survey (see page 38) highlighted the fact that areas which were drained about 25-30 years ago by straightening the river channel and removing the substrate have very low densities of salmonids compared to undrained areas. From 1983 to the present, drainage "maintenance" schemes being undertaken in certain areas to return the river to the original design specifications were identified for investigation. The schemes involved bank and bed works to remove vegetation and shoals of silt and gravel. A study was initiated into various measures recommended for the rehabilitation of salmonid habitat in such drained areas as a fundamental part of these maintenance schemes. There were two experimental approaches.

General recommendations

Recommendations for fisheries measures were agreed in general terms for one tributary (the Stracam) and included re-stoning of the shallow areas to provide nursery habitat for juvenile salmonids. The amount and type of re-stoning was investigated, and this was correlated to the salmonid population densities pre- and post-restocking. Substrate and depth measurements were made at each site and fish population densities determined by quantitative electrofishing at six sites - two of which were controls, where bed works had not been carried out. The results prior to stocking indicated that the mean stone cover in drained and supposedly re-stoned sites was highly significantly lower (p < 0.001) than in the two sites where bed works had not been carried out. There was a highly significant correlation of total fish densities to the proportion of the substrate comprised of stones >10 cm diameter at each site (p < 0.001). Additionally, following restocking the mean survival of stocked salmon fry was over four times greater in the control sites (mean density 33.8 100 m^{-2}) than in the drained sites (mean density 7.5 100 m^{-2}).

These results indicated that re-stoning of the river substrate with stones > 10 cm diameter to cover at least 50% of the area is vital for the rehabilitation of salmonid nursery habitat following river drainage - and that in the case of the R. Stracam this had not been carried out (Kennedy 1984b;

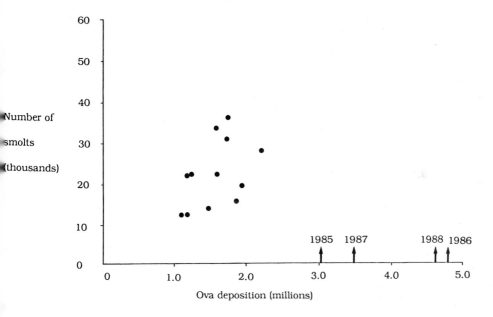

Fig. 1. Stock/recruitment relationship for the R. Bush, showing fully recruited generations (circles) and brood years currently recruiting.

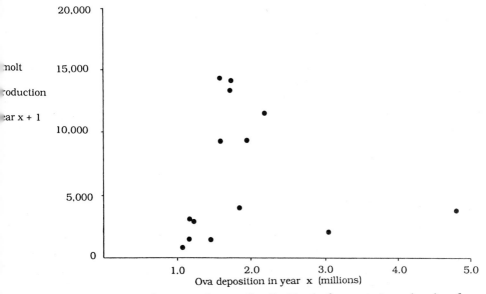

Fig. 2. Production of 1+ smolts in the R. Bush from various levels of ova deposition.

Kennedy and Johnston 1986).

Specific recommendations

In a separate area of the R. Bush, which is subject to a maintenance scheme, recommendations for fisheries rehabilitation measures have been agreed in very specific terms for two stretches. These include the building of groynes to provide deeper water for older fish and substrate re-stoning with stones of 10-40 cm diameter to cover > 50% of the bed in nursery areas. This work is being implemented at present. Surveys to determine fish densities and habitat specifications prior to modification have been carried out, and follow-up investigations will be undertaken in 1989 and 1990.

The freshwater stock/recruitment relationship of Atlantic salmon in the R. Bush

This project is examining how the number of migrating smolts (recruits) varies in relation to number of adult spawners allowed upstream (stock). Also the range of survival from ova to smolt for any given level of stock is being examined in relation to natural environmental variation.

Ova deposition is taken as the index of stock and is derived from the sex ratio of adults released upstream from the traps and their fecundity as measured from females retained for broodstock. This information, together with wild smolt counts, is available from 1973 and continues to be collected. The relationship between stock and recruitment obtained from the range of ova depositions examined to date is given in Fig. 1, together with an indication of the recent higher levels of ova deposition from which the smolt production has not yet fully recruited (owing to smolt migration taking place at 1+, 2+ or 3+ years of age). As yet there is no significant correlation between ova deposition and total smolt production from the fully recruited year classes ($r = 0.461$, $p > 0.1$), though there is a positive trend for increased numbers at higher ova depositions. This may be expected because of the relatively narrow range of ova depositions tested to date (1.07-2.18 million). However, recent higher ova depositions up to 4.79 million (as a result of stock enhancement see page 38) will test whether recruitment is limited or even reduced at much higher stock levels (i.e. is the relationship asymptotic or dome-shaped?).

An indication of possible trends at higher stock levels can be derived from examination of the production of 1+ smolts at various ova depositions (Fig. 2), thus allowing inclusion of cohorts not yet fully recruited. This suggests that production of 1+ smolts is reduced at higher stock levels (above 3 million ova). This may be due to deferred smoltification to older ages as a result of competition reducing growth rates, but this is unlikely given that 1+ smolt production is significantly positively correlated to 2+ smolt production (Fig. 3, $r = 0.634$, $p < 0.05$). In other words, any reduction in numbers of 1+ smolts is not resulting in an increase in 2+ smolts.

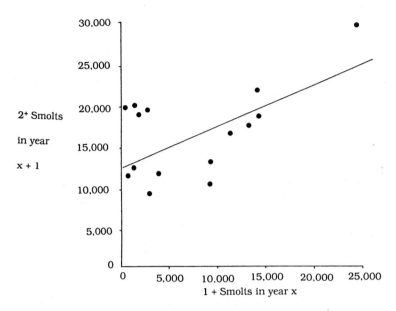

Fig. 3. Relationship between 1+ and 2+ smolt production in the R. Bush.

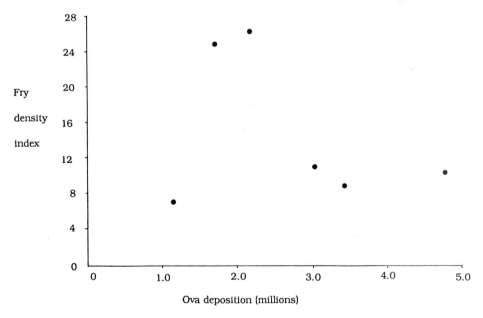

Fig. 4. Average summerling density in the R. Bush related to ova deposition in the previous spawning season.

Support for this hypothesis is provided by analysis of electrofishing survey data on fry densities throughout the Bush system. This has been carried out since 1983 as part of an evaluation of the survival of fry stocked as part of the enhancement project, and involves semi-quantitative backpack electrofishing at a large number of sites during summer (see page 38). Figure 4 illustrates an index of average summerling density in the whole river, related to the ova deposition in the previous spawning season. The trend for lower fry densities at higher ova deposition mirrors the trend in 1+ smolt production described earlier. Therefore, reductions in smolt numbers at higher stock levels seem to reflect mortality rather than growth factors. Whether this represents direct mortality owing to higher stock levels or is related to abiotic events is not yet clear.

The factors regulating freshwater mortality are undoubtedly variable and complex. The wide variation in ova to smolt survival observed from the start of the project (0.69% - 2.13%, mean 1.40% - see Fig. 5) has not yet been attributed to any one factor. Abiotic factors include climatic and other variables, including rainfall patterns affecting distribution of spawners over the system, annual temperature variations, sporadic pollution kills, flood scouring of redds etc. The likely effects of abiotic factors in determining smolt production from any given ova deposition can be judged from the spread of points in Fig. 1 where, for example, ova depositions of 1.16 and 1.17 million gave rise to 12,440 and 21,910 smolts respectively. The range of smolt production at given ova depositions will enable estimation of the buffering needed for environmental variation when setting spawning targets to achieve adequate recruitment.

Biotic factors could include competition for spawning sites and competition (both intra- and inter-specific) for food and space, and predation. The latter has been shown to be potentially important in affecting recruitment from the Bush, Kennedy and Greer (1988) having demonstrated that predation by cormorants *Phalacrocorax carbo* L. may have led to losses of between 51-66% of the potential wild smolt run in 1986, when up to 264 birds were feeding on the river during that period. The results outlined above, indicating that mortality is relatively constant after the parr stage, suggest that cormorant predation on smolts has been fairly constant from year to year. However, Kennedy and Greer (1988) present evidence that this predation may be higher in recent years than in the past. With a predation rate of this magnitude any long-term variation will have a considerable impact on apparent stock/recruitment relationships.

Although the stock/recruitment relationship for the R. Bush is not yet resolved, it is clear that the major factors determining freshwater survival operate early in the freshwater phase. Evidence for this is provided from the positive relationship between 1+ and 2+ smolt production (outlined above) and by a significant positive correlation between average fry density in the river and 1+ smolt production ($r = 0.951$, $p < 0.05$). Thus, mortality seems relatively constant and predictable between the summerling stage and subsequent smoltification at whatever age.

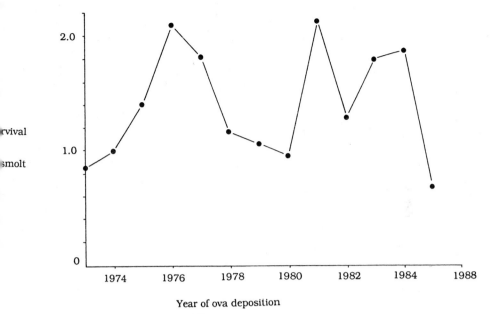

Fig. 5. Ova to smolt survival rate.

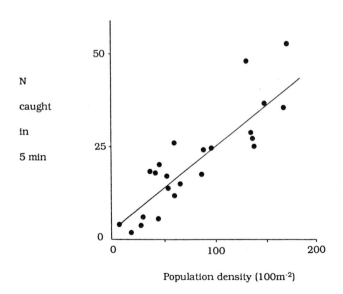

Fig. 6. Calibration graph showing relationship between fry population density in juvenile habitat and number of fry caught per 5 min fishing.

Enhancement of Atlantic salmon stocks in the R. Bush

The aims of this work were twofold: firstly to investigate the efficacy of juvenile salmon stocking as a technique for improving adult runs, and secondly in relation to the River Bush to generate high ova deposition rates for the stock/recruitment study.

The whole river was surveyed and stretches categorised as suitable for stocking on the basis of habitat type and lack of natural spawning. Fry stocking densities of between 2 m^{-2} and 5 m^{-2} were used, depending on habitat type and the totals for each area calculated. The progeny of wild x wild crosses at the Bush hatchery were transported as swim-up fry to the stocking area in polythene bags with oxygen. Stocking was carried out using watering cans to achieve an even distribution of fry over the available habitat. Half a million fry were stocked in both 1983 and 1984; however, only 0.19 million fry were available for stocking in 1985 and 0.25 million in 1986.

Electrofishing of up to 150 sites in both stocked and unstocked areas (controls) was carried out annually using a specially developed semi-quantitative technique. A single anode backpack is used to sample fry over 5 min periods. This technique has been calibrated by sampling within stop-netted sections prior to full quantitative electrofishing and a high degree of correlation obtained between fry caught in 5 min and the actual population density (r = 0.87, p < 0.001, Fig. 6). This provided a good measure of the survival of stocked salmon to the summerling stage and allowed comparison with naturally spawned areas of the river. It is thus possible to identify areas giving consistently good survival and other areas where poor survival indicated water quality or other problems.

It was originally expected that the success of all river enhancement stocking would best be quantified by monitoring smolt counts at the trap at Bushmills. However, the complex relationship emerging between stock and recruitment together with cormorant predation has meant that a direct relationship between numbers stocked and extra smolts produced has not been apparent.

The success of the enhancement programme can, however, be judged in several ways. The electrofishing surveys carried out every summer following stocking indicated good survival of stocked fry, equivalent to that found in controlled stocking experiments at Altnahinch. Thus, the stocked fish were surviving beyond the critical early mortality period (discussed above, page 36) and would be expected to contribute to smolt production. Electrofishing surveys carried out during the years of stocking, and subsequently, to monitor the distribution of natural spawners in the system have indicated that areas of the river now utilised by natural spawners had no spawning before the enhancement programme. This includes extensive areas of head-waters that are now contributing significantly in overall terms to the river production. It is tempting to suggest that these spawners are the progeny of adults produced from the enhancement programme. Finally, the numbers of adult fish entering the Bush to spawn has increased since enhancement (reversing the previous trend - see Fig. 7), although this increase is also partly attributable to a larger component of hatchery-reared fish produced to service the

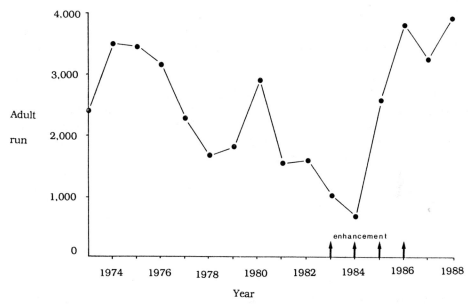

Fig. 7. Adult wild salmon runs on the R. Bush, 1973-1988. The years during which enhancement stocking was carried out are indicated (arrows).

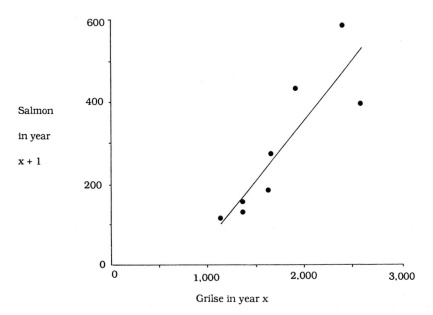

Fig. 8. Relationship between grilse run in year x and subsequent salmon run in years x + 1 on the R. Bush.

microtagging programme (see below). Enhancement stocking has not been carried out since 1986, as adult runs are no longer declining, but has been retained as a management option.

The marine survival of R. Bush salmon stocks

This project aims to investigate the factors influencing the survival at sea of salmon smolts migrating from the R. Bush until their return as adult salmon. Central to this is the investigation of migration patterns and quantification of exploitation levels in high seas and home waters fisheries. Survival from smolt to returning adult has been monitored annually from the smolt counts and trapped adult numbers. Angling returns from the section of river downstream of the adult trap are added to trap counts to derive total river returns. Age analysis is used to separate overlapping cohorts and determine sea age of the adults.

Marine survival of wild smolts

The annual survival rates for each smolt release year from 1974-1986 are given in Table 1. Overall survival of wild salmon to the river during this period ranged from 6.26-12.08% (mean 8.61%). The return rate was fairly constant, with the high marine survival from the 1979, 1981 and 1986 smolt releases being the exceptions. The marine survival rates of 1SW and 2SW fish were significantly correlated throughout the period (r = 0.877; p < 0.01), suggesting that the critical regulating factor operated at the smolt or early post-smolt stage, and was not related to differences in exploitation patterns. This relationship permits the prediction of 2SW runs from grilse numbers in the previous year (Fig. 8). The relative stability of marine survival results in a significant positive correlation between smolt release and adult return (r = 0.815; p < 0.01; Fig. 9). This provides a reasonably accurate basis from which to predict the likely adult run from a given smolt release (assuming relative constancy of fishing mortality).

Hatchery smolt production and marine survival

Hatchery smolts were produced for the original project aim of investigating ranching of artificially produced salmon. However, the problems with broodstock losses and low survivals from egg to released smolt hindered the development of this aspect of the research for some years (see Kennedy and Johnston 1986 for details). Significant improvements in hatchery management have contributed to better survival of broodstock and increased survival from hen to smolt, leading to more smolts per hen retained. Hatchery-reared smolt output is currently in the region of 20-25,000 per year which yields between 650-900 adults to the river. The development of a self-sustaining hatchery stock is a consequence of improved rearing techniques acquired through experience. With this run now servicing the microtagging programme, as well as contributing to angling, the role of "ranched" salmon in the project is now significant. The marine survival of hatchery-reared

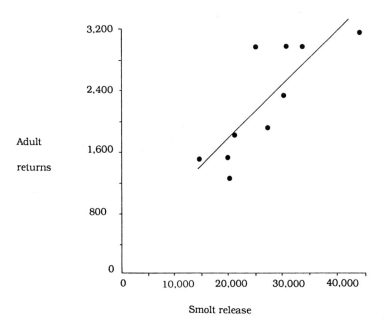

Fig. 9. Wild smolt production and adult returns to the river.

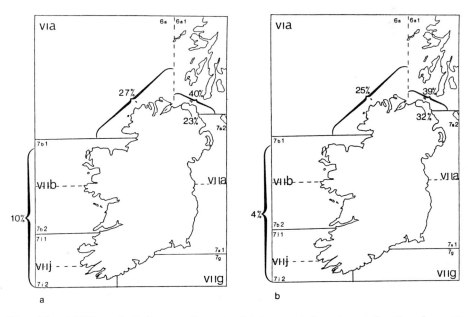

Fig. 10. 1987 exploitation pattern in Irish coastal waters of grilse from the 1986 hatchery-reared (a) and wild (b) smolt releases from the R. Bush. Fishery areas follow ICES designations.

smolts over the period 1973-1987 has been low and variable (Table 2), with average survival being 5.06 times greater in wild smolts over the same period. Recent returns have indicated improving survival, with survival to adult return of the 1985 smolt release having exceeded 4%.

Exploitation patterns and levels

Assessment of the exploitation patterns of adults at sea was originally undertaken using Carlin tagging of smolts and floy tagging of broodstock kelts. Although these releases gave some indication of the migration routes of salmon from the Bush, they could not be used to quantify exploitation in various fisheries owing to biased reporting of tags.

A programme of microtagging of hatchery-reared smolts was introduced in 1983 and this was extended to include wild smolts since 1986. A total of 94,739 microtagged smolts have been released to date, with 4,718 recaptures being reported. This work forms part of an international programme of microtagging carried out under the auspices of ICES, involving nine salmon producing countries. National monitoring programmes are in place to examine home water and high seas fisheries and return recovered tags to the country of origin, together with biological details of the captured fish.

Using information obtained from the microtagging programme it is possible to estimate the survival of tagged smolts back to the Irish coast and to examine the level and patterns of exploitation by various fisheries. These estimates are derived using known counts of tagged adults returning to the river and the number of tags returned from the various fisheries. Numbers likely to have been taken in fisheries are estimated by scaling up tag numbers from the sampled catch to the declared catch and by further expansion encompassing estimates of unreported catch.

In Table 3 exploitation and survival data are given for grilse returns from microtagged R. Bush salmon returning as grilse to home waters in 1987. Estimated levels of marine exploitation for returns from hatchery-reared smolts were high, ranging from 77.5% (2+ smolts) to 93.9% (1+ smolts). This compares well with the range of exploitation for similar releases in the 1986 fishery (75%-82%; Crozier and Kennedy 1987). The estimated exploitation rate of the 1986 release of microtagged wild R. Bush smolts is also given. This is the first exploitation estimate available from trapping and tagging of wild smolts in an Irish river and suggests that they were subject to lower fishing mortality than hatchery releases from the same river. The actual percentage return to the river is also given for these three tagged groups, together with estimated returns to the Irish coast. The tagged wild smolts showed a much better return than the hatchery-reared fish, with an estimated 17.5% of wild salmon returning as grilse to the Irish coast prior to exploitation. Correcting this figure for tagging mortality (available by comparing tagged and untagged wild salmon returning to the Bush), it is possible to estimate the coastal return of untagged wild salmon at 26.2%.

Table 1. Marine survival data for R. Bush wild smolts.

Year of release	Number of smolts	% Return as adults to R. Bush		
		1SW	2SW	Total
1974	43,958	6.75	0.47	7.22
1975	33,365	8.52	0.45	8.97
1976	21,021	7.76	0.90	8.66
1977	19,693	6.94	0.80	7.74
1978	27,104	6.13	1.00	7.13
1979	24,733	10.48	1.60	12.08
1980	20,139	5.67	0.59	6.26
1981	14,509	9.46	0.92	10.38
1982	10,694	7.79	-	- *
1983	26,804	-	1.69	-
1984	30,009	6.40	1.45	7.85
1985	30,518	7.87	1.92	9.79
1986	18,442	9.75	1.94	11.70

* 1984 adult run was a partial count owing to electric barrier breakdown.

Table 2. Marine survival of R. Bush hatchery-reared smolts.

Year of release	Number released	% Return as adult to R. Bush
1973	275	2.50
1974	4,349	1.56
1975	7,065	0.32
1976	37,370	1.19
1977	10,577	0.27
1978	10,642	0.26
1979	10,631	0.35
1980	4,841	0.80
1981	2,702	1.63
1982	1,338	0.75
1983	4,754	0.40
1984	2,711	0.88
1985	17,966	4,19
1986	25,159	2.10
1987	20,639	3.80*

* Grilse returns only.

Table 3. Marine exploitation and survival of the 1986 release of hatchery-reared and wild smolts from the R. Bush.

Source	Age	Number released	Number recovered	Marine exploitation (%)	Coastal return (%)	R: re (%
Hatchery	1+	2,312	25	93.9	2.0	0.
"	2+	21,847	1,198	77.5	9.7	2.
Wild run	1+/2+	1,161	129	68.5	17.5	5.

Figure 10 illustrates the grilse exploitation pattern in Irish coastal waters in 1987 of hatchery-reared (2+ at release) and wild smolts from the R. Bush, based on 23,008 tagged releases and 1,327 recaptures. An estimated 37% of returning hatchery-reared Bush fish were taken in the Republic of Ireland Fishery, mostly in the Donegal area. The comparable figure for the tagged wild Bush fish is lower at 29%, but is still significant in proportion to the 39% of this group taken by the N. Ireland fishery. This research has revealed and quantified previously unknown exploitation in west and south-west Ireland of salmon returning to a N. Ireland river, with up to 52.7% of returning hatchery-reared fish being taken in the Republic of Ireland in 1986.

Information on timing and routes of migration of returning adults is also becoming available through this programme. A high proportion of tagged recaptures of Bush fish are taken in nets to the east of the river (51.3% in 1987), suggesting a run of returning adults south along the west coast of Scotland, prior to a westerly movement towards the river. Unfortunately with no tag recovery programme in this area it is not possible to confirm this or to quantify the exploitation of Bush (or any other Irish) stocks in this fishery.

Information on adult exploitation rates is of importance for DANI in responding to a new approach in managing salmon by assessing national catches with a view to national management. ICES is presently working towards a system of high seas and home water quotas which will ultimately be used as a means of ensuring adequate spawning escapement (Anon. 1988). The role of the R. Bush and other index rivers will be central to this approach to management.

CONCLUSIONS

Assessment, management and rehabilitation information resulting directly from River Bush research has been incorporated into programmes carried out by DANI Drainage Division, Foyle Fisheries Commission and the Fisheries Conservancy Board for Northern Ireland. Angling clubs and fishermen's associations have also benefited from management and development advice available as a result of the early stages of this research. For example, information on marine survival and exploitation has been used to advise on requested changes in legislation relating to seasonal and weekly closed periods for commercial salmon fishing in N. Ireland.

Research on the Bush is a series of interrelated short- and long-term projects, designed to examine in depth many aspects of salmon ecology and management. Many of the shorter-term aspects have been completed (and results published) and these are providing management information as well as identifying further areas of research. However, a major part of the research is structured on a long-term basis. This is a consequence of the long life cycle of the Atlantic salmon, the need to monitor specific cohorts over time and the effects of environmental variation on natural mortality, migration patterns, timing and exploitation.

Further research areas have been identified which will address the factors affecting survival during two critical phases of the salmon life cycle: early juvenile development, and the post-smolt period. Research will be carried out to identify the major abiotic factors influencing recruitment during the freshwater phase. This will examine climatic and other variables thought to account for the spread of recruitment from given levels of ova deposition. Mortality factors during the early post-smolt period will be investigated, particular attention being given to multi-species predation on smolts and whether food availability influences post-smolt survival.

ACKNOWLEDGEMENTS

Thanks are due to Mr K.U. Vickers for his hard work in establishing the Department's facilities at the R. Bush, to the previous hatchery managers, Mr J. Greer, Mr D. Anderson and Dr P. Johnston, to our colleagues Mr C.D. Strange and Mr B. Hart, and all the other workers who have been involved in the project since its inception.

REFERENCES

ANON. 1973 River Bush Project. *Department of Agriculture for Northern Ireland Annual Report on Research and Technical Work, 1973*, 121-8.

ANON. 1988 Report of the working group on north Atlantic salmon. *ICES C.M. 1988*, Assess: 16 (mimeo).

CROZIER, W.W. and KENNEDY, G.J.A. 1987 Marine survival and exploitation of R. Bush hatchery salmon (*Salmo salar* L.) as assessed by microtag returns to 1986. *ICES C.M. 1987/39* (mimeo).

CROZIER, W.W. and MOFFETT, I.J.J. 1989 Amount and distribution of biochemical-genetic variation among wild populations and a hatchery stock of Atlantic salmon *Salmo salar* L., from north-east Ireland. *Journal of Fish Biology* 35, 665-77.

KENNEDY, G.J.A. 1981a The reliability of quantitative juvenile salmon estimates using electro-fishing techniques. Proceedings of the Atlantic Salmon Trust Workshop, Windermere, 1981. (Unpublished working paper.)

KENNEDY, G.J.A. 1981b Some observations on the interrelationships of juvenile salmon (*Salmo salar* L.) and trout (*Salmo trutta* L.). *Proceedings 2nd British Freshwater Fisheries Conference 1981*, 143-9.

KENNEDY, G.J.A. 1984a Factors affecting the survival and distribution of salmon (*Salmo salar* L.) stocked in upland trout (*Salmo trutta* L.) streams in Northern Ireland. *Symposium on Stock Enhancement in the Management of Freshwater Fisheries, EIFAC, Budapest 1982,* EIFAC Tech. Pap. (42), Suppl. Vol. 1, 227-42.

KENNEDY, G.J.A. 1984b The ecology of salmonid habitat re-instatement following river drainage schemes. *Institute of Fisheries Management (N.I. Branch) Study Course,* 1-13. Institute of Continuing Education, University of Ulster.

KENNEDY, G.J.A. 1985 River pollution - How much does it cost fisheries? Unpublished paper, presented at DANI Advisers and Lecturers Conference, Loughry College of Agriculture and Food Technology, 1985.

KENNEDY, G.J.A. 1986 Silage effluent pollution - costs and prevention. *Agriculture in Northern Ireland* 60 (12), 402-6.

KENNEDY, G.J.A. 1988 Stock enhancement of Atlantic salmon (*Salmo salar* L.). *Proceedings of the Third International Atlantic Salmon Symposium Biarritz,* 345-88. London. Croom Helm.

KENNEDY, G.J.A. (in prep.) *Evaluation of production from salmonid nursery habitat in the British Isles.* Atlantic Salmon Trust Special Publication.

KENNEDY, G.J.A. and GREER, J.E. 1988 Predation by cormorants (*Phalacrocorax carbo* L.) on the salmonid populations of the River Bush. *Aquaculture and Fisheries Management* 19, 159-70.

KENNEDY, G.J.A. and JOHNSTON, P.M. 1986 A review of salmon (*Salmo salar* L.) research on the River Bush. In W.W. Crozier (ed.), *Proceedings of the Institute of Fisheries Management (NI Branch), 17th Annual Study Course,* 49-69. University of Ulster, Coleraine.

KENNEDY, G.J.A. and STRANGE, C.D. 1980 Population changes after two years of salmon (*Salmo salar* L.) stocking in upland trout (*Salmo trutta* L.) streams. *Journal of Fish Biology* 17, 577-86.

KENNEDY, G.J.A. and STRANGE, C.D. 1981a Efficiency of electric fishing for salmonids in relation to river width. *Fisheries Management* 12, 55-60.

KENNEDY, G.J.A. and STRANGE, C.D. 1981b Comparative survival from salmon (*Salmo salar* L.) stocking with eyed and green ova in an upland stream. *Fisheries Management* 12, 43-8.

KENNEDY, G.J.A. and STRANGE, C.D. 1986a The effects of intra and interspecific competition on the survival and growth of stocked juvenile Atlantic salmon (*Salmo salar* L.) and resident trout (*Salmo trutta* L.) in an upland stream. *Journal of Fish Biology* 28, 479-89.

KENNEDY, G.J.A. and STRANGE, C.D. 1986b The effects of intra and interspecific competition on the distribution of stocked juvenile Atlantic salmon (*Salmo salar* L.) in relation to depth and gradient in an upland trout (*Salmo trutta* L.) stream. *Journal of Fish Biology* 229, 199-214.

In: Steer, M.W. (ed.) 1991 *Irish Rivers : Biology and Management*, pp 47-57.
Royal Irish Academy, Dublin.

THE DIEL FEEDING PATTERN OF BROWN TROUT
(*SALMO TRUTTA* L.) IN ACID STREAMS DURING
THE SUMMER AND AUTUMN MONTHS

Mary Kelly-Quinn
Zoology Department, Trinity College Dublin

ABSTRACT

A brown trout (*Salmo trutta* L.) production study in tributary streams of the Avonmore River in Co. Wicklow is currently in progress (June 1987 to May 1989). This present paper is part of an investigation into the factors affecting production in these acid waters. Stomach samples were collected every three hours over five periods from May to October 1988. The diel feeding pattern of all age groups was characterised by a morning and evening peak in feeding intensity in all months. A further afternoon increase in feeding was noted for August and September. Feeding intensity was highest in June and August. A sharp decline in morning feeding was noted for October. There was considerable diel variation in the composition of the diet. In general, aquatic invertebrates dominated the diet during the hours of darkness and early morning. Terrestrial invertebrates were important in the afternoon and well into the evening. Samples collected during this study will be later used to estimate the daily ration of three age groups of brown trout.

INTRODUCTION

The study of brown trout production in tributary streams of the Avonmore River in Co. Wicklow commenced in June 1987 and will continue until May 1989. This acid water (pH 5.5-6.5) system is characterised by slow-growing (L_1 = 6.0) population of trout (Kelly-Quinn in preparation). Investigation of the factors affecting production in this system includes food consumption. The data presented in this paper form part of the information collected to estimate daily ration. This aspect of the work will continue until April 1989.

Extensive studies on the diel feeding of field-caught salmonids other than brown trout have been carried out (Shimazaki and Mishima 1969; Bisson

1978; Narver 1970; Doble and Eggers 1978; Godin 1981; Sagar and Glova 1988). The limited number of studies dealing with brown trout include Hoar (1942), Chaston (1969) and Elliott (1970b). This present study is based on a more extensive sampling programme involving larger numbers of brown trout in three different age groups. The findings will add to existing knowledge on the diet of brown trout in Irish rivers. The management implications of the findings are discussed.

STUDY AREA

The diel feeding study was carried out on trout in the Glenealo River which flows through the Valley of Glendalough (Fig. 1). It flows in a west-east direction and joins the Glendassan River, a tributary of the Avonmore River. Sampling was carried out on the stretch of the Glenealo River which flows between the Upper and Lower lakes. Here the river consists of short riffle-pool sequences, with an average width of 4.0 m and average depth of 0.18 m. The substratum consists of small stones embedded in sand, gravel and occasional boulders.

The geology of the area consists of granite at the head of the valley and mica schist in the rest of the catchment. The floor of the valley is covered with alluvial peat. Bank vegetation consists largely of *Betula* sp. (birch), *Quercus* sp. (oak), *Salix* sp. (willow), *Alnus glutinosa* (alder), *Pinus sylvestris* (Scots pine) with occasional *Acer pseudoplatanus* (sycamore) and *Fraxinus excelsior* (ash), together with *Ulex* sp. (gorse) and a variety of grass species.

MATERIALS AND METHODS

The study area was divided into eight similar 30 m stretches. Twenty to twenty-five trout were collected every 3 hours for 24 hours using electrofishing equipment. Length (fork) was recorded in cm and the stomach contents were removed by flushing using Foster's (1977) modification of the Seaburg (1957) technique. The fish were fin-clipped to allow identification of fish whose stomachs had previously been flushed. The samples were deep-frozen for later analysis.

The trout were separated into three age groups, 0+ (< 7.9 cm), 1+ (4.5-12.5 cm) and 2+/3+ (10.0-21.8 cm). Table 1 gives the numbers of fish from each age group which were taken during 24-hour sampling periods. Sampling was carried out in May (17/18), June (26/27), August (4/5), September (10/11) and October (8/9) 1988. No sampling was carried out in July because of high water levels. The 0+ fish, because of their small size, were not stomach-flushed until August.

Stomach contents were identified to the lowest taxon. The taxa were divided into twelve broad categories and expressed in terms of percentage numerical composition. The data were also combined into three broad categories: terrestrial invertebrates, adult and emerging insects of aquatic origin, and other aquatic fauna. Feeding intensity was expressed in terms of average numbers of fresh prey items per stomach and average wet weight (mg) of entire contents.

49

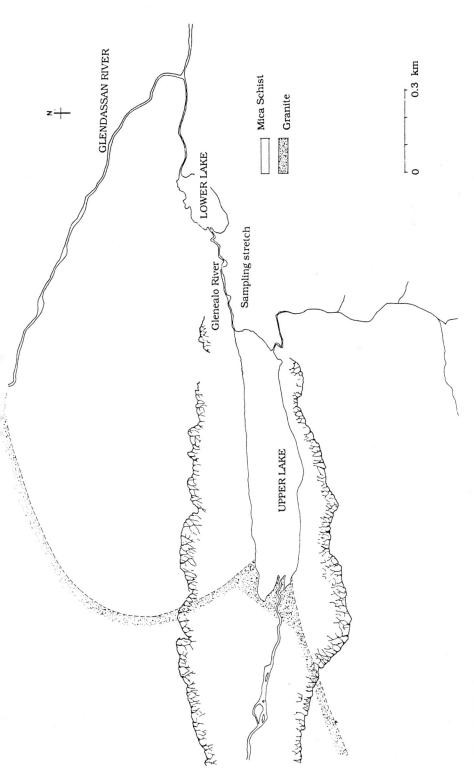

Fig. 1. Study area.

50

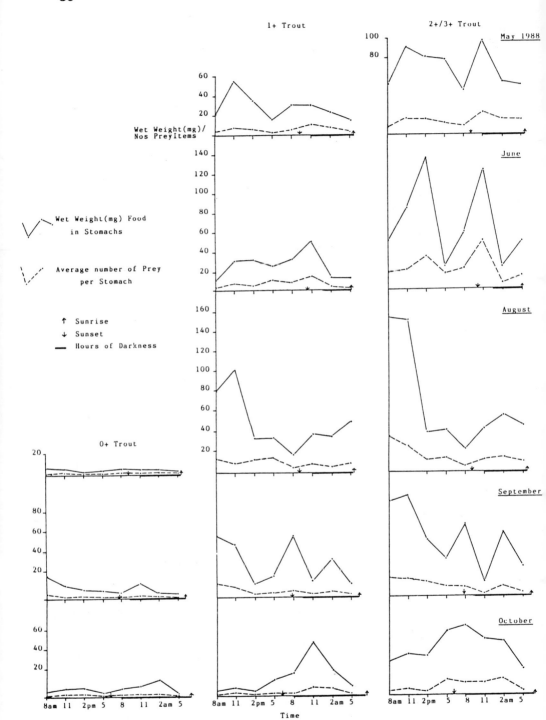

Fig. 2. Diel variation in the biomass (wet weight mg) and numbers of prey in the stomachs of 0+, 1+ and 2+/3+ trout.

Table 1. Numbers of 0+, 1+ and 2+/3+ trout stomachs examined at each sampling period.

Date (Month)

Time	May 0+	1+	2+/3+	June 0+	1+	2+/3+	August 0+	1+	2+/3+	Sept. 0+	1+	2+/3+	October 0+	1+	2+/3+
8 am	0	13	9	0	9	9	7	7	13	6	7	9	7	6	7
11am	0	13	9	0	10	10	6	10	9	6	8	8	6	8	7
2 pm	0	12	10	0	9	9	6	7	6	8	7	6	7	9	7
5 pm	0	9	14	0	10	10	6	8	8	6	7	7	7	7	6
8 pm	0	11	9	0	14	8	6	9	6	6	8	6	6	8	8
11pm	0	10	10	0	9	13	7	8	6	6	9	8	6	8	8
2 am	0	9	12	0	11	11	6	8	8	7	8	8	6	9	8
5 am	0	12	8	0	10	10	8	9	6	6	10	7	6	9	8

RESULTS

Feeding intensity

Food was present in the stomachs at all sampling times over each 24-hour period. Only 44 (5.1%) of the 860 stomachs were empty. There was no significant ($X^2 = 0.57$, $P > 0.01$) difference in the number of empty stomachs recorded in daylight or darkness. The number of empty stomachs recorded in October (21) was significantly ($X^2 = 25.95$, $P < 0.01$) higher than that recorded in May (9), June (2), August (2) or September (10).

In terms of the amount of food in the stomachs there was significant ($H = 26.85$, $P < 0.01$, Kruskal-Wallis one-way ANOVA) diel variation. Feeding intensity, as indicated by the number of fresh items per stomach, increased from early morning to a maximum between 8 am and 2 pm. A second peak occurred in the evening between 11 pm and 2 am (Fig. 2). In August and September a further increase in feeding intensity was observed in the afternoon. Biomass figures also confirmed these findings. This pattern of feeding

52

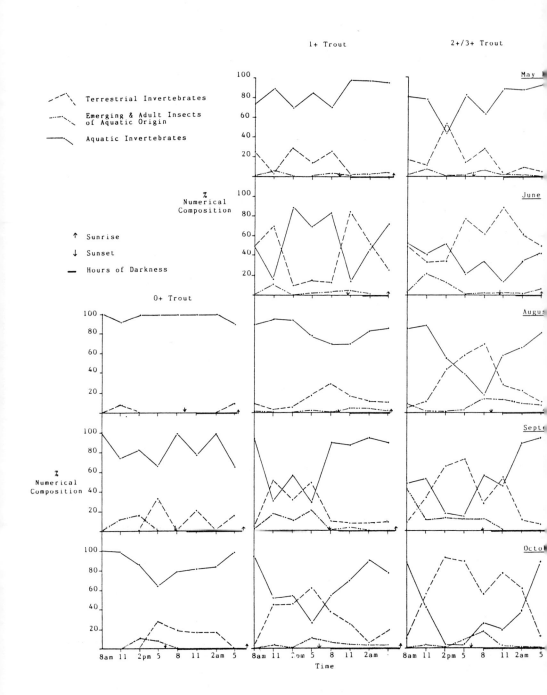

Fig. 3. Diel variation in the % numerical composition of the diet of 0+, 1+ and 2+/3+ trout.

activity was essentially the same for each of the three age groups.

In May and June feeding intensity, in terms of numbers of prey items per stomach, was highest in the evening, yet on a number of occasions biomass figures were lower (Fig. 2). The discrepancies may be explained by the size of the prey taken. In the mornings the diet was dominated by large (8-12 mm) trichopteran larvae which were on average 4 to 6 times heavier than adult dipterans which formed the bulk of the evening diet. In August and September the number of food items per stomach was only slightly higher at the morning peak but again the greater weight of the aquatic prey items gave higher biomass figures than for the evening period. The discrepancy did not exist for October as only small (1-3 mm) trichopteran larvae were consumed. Having emerged as adults in August and September large trichopteran larvae were relatively unavailable (Kelly-Quinn in preparation).

There was also a significant (H = 22.67, P < 0.01, Kruskal-Wallis one-way ANOVA) difference in feeding intensity between months. The largest numbers and biomass of food items were recorded in June and August. Figures for May and September were similar. By October early-morning feeding was greatly reduced (Fig. 2).

With regard to differences in the amount of food in the stomachs of different age groups there was, as expected, a larger amount of food in the stomachs of the larger fish.

Composition of the diet

Kelly-Quinn (in preparation) describes the general diet of trout in the Glenealo/Glendassan/Glenmasnass rivers, which is dominated by Plecoptera (*Protonemura meyeri* Pict, *Amphinemura sulicollis* Steph. and *Leuctra* spp), Trichoptera (*Hydropsyche siltalai* Dohler, *H. pellucidula* Curtis and *Rhyacophila dorsalis* Curtis), terrestrial invertebrates and to a lesser extent chironomid larvae. Results of the 24-hour feeding analysis generally agreed with this result. However, there was considerable diel variation in the composition of the diet both from month to month and between age groups. The 0+ trout had a largely aquatic diet which was dominated by trichopteran larvae, plecopteran nymphs, chironomid and simuliid larvae (Fig. 3). Adult terrestrial insects were taken in small numbers (< 35% of food intake) in the morning and evening in August and in the early afternoon during September and October. Emerging adult plecopterans and ephemeropterans were also taken in the early afternoon during September and October.

The diet of 1+ trout was dominated by aquatic food during May, August, September and October (Fig. 3). These were most important during the hours of darkness and early morning. Terrestrial invertebrates did, however, contribute up to 60% of the diet from midday to early evening. In June these terrestrial fauna formed up to 80% of the food intake during both the morning and evening feeding peaks. Midday to early evening was dominated by aquatic food, mainly Trichoptera and Plecoptera. As was the case with 0+ fish, emerging and adult aquatic insects comprised up to 20% of the midday diet. Small numbers were also taken during the mornings and evenings of the other months.

With regard to 2+/3+ fish, aquatic invertebrates dominated the diet in May, except at 2 pm, when adult dipterans were taken in large numbers. For the remaining months the diet of these fish was largely food of terrestrial origin (Fig. 3), which comprised as high as 90% of the diet at any one sampling time. Aquatic fauna were, however, important during the hours of darkness or early morning. Small numbers of trout (0+) and minnow were consumed by both 1+ and 2+/3+ fish but these were mainly taken at dusk or during the hours of darkness.

DISCUSSION

The results of this study suggest that during the summer and autumn months brown trout fed continuously over the 24-hour cycle. However, their diel feeding activity was characterised by two main peaks in feeding intensity. Morning feeding begins shortly after sunrise with maximum stomach fullness achieved between 8 am and 2 pm. The second peak occurred in the evening just after sunset. Feeding intensity remained high through the twilight period and well into the dark from August to October. This feeding pattern is similar to that reported by Hoar (1942), Shimazaki and Mishima (1969) and Godin (1981). A further afternoon increase in feeding activity also occurred in September and August. A similar feature was noted by Elliott (1970b) and Kelly-Quinn and Bracken (1990).

Light intensity probably plays an important role in determining the type of food consumed at any time over the 24 hours. Activity of terrestrial insects during the hours of daylight increases their chances of becoming part of the drift. This was reflected in the high afternoon occurrence of terrestrial fauna in the diet of all age groups. Several authors have also reported a high daytime occurrence of terrestrial invertebrates in the diet of trout and also in the drift (Elliott 1967a; 1967b; 1970b; Chaston 1969; Kelly-Quinn 1980). Emerging and adult insects of aquatic origin also characterised the daylight hours. With the exception of 0+ trout, which had a predominantly aquatic diet at all times of the day, aquatic organisms formed the largest proportion of the food intake during the hours of darkness or early morning. It is generally accepted that during these hours there is increased activity of some invertebrates on the substratum (Elliott 1968; 1970a) or an increase in their numbers in the drift (Waters 1962; 1965; Elliott 1965; 1967a; 1967b; Allan 1978). Trout fry and minnow were generally taken at dusk or during the hours of darkness. Downstream movement of fry during the early hours of the night may make them more susceptible to predation (Elliott 1966). Variation in the type of food consumed with time of day highlights the importance of stating the time of sampling in any report dealing with the diet of trout.

Apart from availability of prey, feeding intensity is affected by water temperature (Elliott 1975). Water temperatures ranging from 11.8° to 18.0°C characterised the May to September period (Table 2). There was a sharp drop in water temperature in October which may have reduced feeding in the morning and early afternoon. Gardiner and Geddes (1980) noted that juvenile salmon moved from feeding stations to shelter at temperatures below

10°C. Elliott (1975) noted a reduction in the number of meals consumed per day with falling water temperature. The unavailability of large trichopteran larvae, a group which predominated in the morning diet, or other aquatic invertebrates must also be considered as a factor affecting feeding intensity.

Information from this study on the biomass of food in the stomachs together with that collected between November 1988 and April 1989 will be used to estimate the daily ration of trout and will be presented in a later paper.

In terms of the management of salmonid rivers this study highlights the often underestimated importance of the terrestrial input to the system, especially in low-productivity waters. The importance of the terrestrial component in the diet of the older fish of spawning age increases when one considers that food consumed during the summer is more important for growth and development, because of the increased metabolism associated with high water temperatures. This, coupled with the fact that the adult insect diet coincides with low availability of some bottom fauna, gives a greater ecological value to the terrestrial input than first anticipated when their contribution is viewed on an annual basis (Hunt 1975; Mann and Orr 1969). Therefore the management, protection and rehabilitation of salmonid streams and rivers must encompass the natural bank vegetation, a source of terrestrial invertebrates.

Table 2. Water temperatures (°C) recorded over 24-hour periods from May to October 1988.

	Time							
	8am	11am	2pm	5pm	8pm	11pm	2am	5am
May	12.0	12.3	12.5	12.9	12.6	12.2	12.0	11.8
June	17.2	17.4	17.8	18.0	18.0	18.0	18.0	17.6
Aug.	14.3	14.6	14.8	15.4	15.2	15.1	14.5	14.2
Sept.	14.2	14.7	14.7	14.4	14.3	14.0	13.8	13.1
Oct.	10.1	10.3	10.5	10.4	10.1	10.1	10.1	9.9

ACKNOWLEDGEMENTS

I wish to thank Dr J.J. Bracken, Zoology Dept., University College Dublin, for reading the manuscript. I am indebted to Mr Noel Quinn and Ms Valerie Cox for help with the field-work. I also wish to acknowledge and thank the Department of Education for funding this study and the Department of Zoology, Trinity College Dublin, for providing the necessary facilities.

REFERENCES

ALLAN, J.D. 1978 Trout predation and the size composition of stream drift. *Limnology and Oceanography* 23 (6), 1231-7.

BISSON, P.A. 1978 Diel food selection by two sizes of rainbow trout (*Salmo gairdneri*) in an experimental stream. *Journal of the Fisheries Research Board of Canada* 35, 971-5.

CHASTON, I. 1969 Seasonal activity and feeding pattern of brown trout (*Salmo trutta*) in a Dartmoor stream in relation to availability of food. *Journal of the Fisheries Research Board of Canada* 26, 2165-71.

DOBLE, B.D. and EGGERS, D.M. 1978 Diel feeding chronology, rate of gastric evacuation, daily ration and prey selectivity in Lake Washington juvenile sockeye salmon (*Oncorhynchus nerka*). *Transactions of the American Fisheries Society* 107, 36-45.

ELLIOTT, J.M. 1965 Daily fluctuations of drift invertebrates in a Dartmoor Stream. *Nature* (London) 205, 1127-9.

ELLIOTT, J.M. 1966 Downstream movement of trout fry in a Dartmoor stream. *Journal of the Fisheries Research Board of Canada* 23 (1), 157-9.

ELLIOTT, J.M. 1967a The food of trout (*Salmo trutta*) in a Dartmoor stream. *Journal of Applied Ecology* 4, 59-71.

ELLIOTT, J.M. 1967b Invertebrate drift in a Dartmoor stream. *Archiv für Hydrobiologie* 63, 202-37.

ELLIOTT, J.M. 1968 The daily activity pattern of mayfly nymphs (Ephemeroptera). *Journal of Zoology* (London) 155, 201-21.

ELLIOTT, J.M. 1970a The diel activity patterns of caddis larvae. *Journal of Zoology* (London) 160, 279-90.

ELLIOTT, J.M. 1970b Diel changes in invertebrate drift and the food of trout, *Salmo trutta* L. *Journal of Fish Biology* 2, 161-5.

ELLIOTT, J.M. 1975 Number of meals in a day, maximum weight of food consumed in a day and maximum rate of feeding for brown trout, *Salmo trutta* L. *Freshwater Biology* 5, 287-303.

FOSTER, J.R. 1977 Pulsed gastric lavage, an efficient method of removing the stomach contents of live fish. *Progressive Fish Culturist* 39, 166-9.

GARDINER, D. and GEDDES, J. 1980 The influence of body composition on the survival of juvenile salmon. *Hydrobiology* 69, 67-72.

GODIN, J.G.J. 1981 Daily pattern of feeding behaviour, daily rations and diets of juvenile pink salmon (*Oncorhynchus gorbuscha*) in two marine bays of British Columbia. *Canadian Journal of Fisheries and Aquatic Sciences* 38, 10-15.

HOAR, W.S. 1942 Diurnal variations in feeding activity of young salmon and trout. *Journal of the Fisheries Research Board of Canada* 6, 90-101.

HUNT, R.L. 1975 Food relations and behaviour of fishes - use of terrestrial invertebrates as food by salmonids. In A.D. Hasler (ed.), *Coupling of land and water systems*, 137-51. New York. Springer-Verlag.

KELLY-QUINN, M. 1980 A study of brown trout (*Salmo trutta* L.) populations in the River Dodder catchment with particular reference to the Owendoher Stream. Unpublished Ph.D. thesis, National University of Ireland.

KELLY-QUINN, M. and BRACKEN, J.J. 1990 A seasonal analysis of the diet and feeding dynamics of brown trout, *Salmo trutta* L., in a small nursery stream. *Aquaculture and Fisheries Management* 21, 107-24.

MANN, R.H.K. and ORR, D.R. 1969 A preliminary study of the feeding relationships in a hard-water and a soft-water stream in southern England. *Journal of Fish Biology* 1, 31-44.

NARVER, D.W. 1970 Diel vertical movements and feeding of underyearling sockeye salmon and the limnetic zooplankton in Barbine Lake, British Columbia. *Journal of the Fisheries Research Board of Canada* 27, 281-316.

SAGAR, P.M. and GLOVA, G.J. 1988 Prey preferences of a riverine population of juvenile chinook salmon, *Oncorhynchus tshawytscha*. *Journal of Fish Biology* 31, 661-73.

SEABURG, K.G. 1957 A stomach sampler for live fish. *Progressive Fish Culturist* 19, 137-9.

SHIMAZAKI, K. and MISHIMA, S. 1969 On the diurnal changes of the feeding activity of salmon in the Okhotsk Sea. *Bulletin Faculty of Fisheries Hokkaido University* 20, 82-93.

WATERS, T.F. 1962 Diurnal periodicity in the drift of stream invertebrates. *Ecology* 43, 316-20.

WATERS, J.M. 1965 Interpretation of invertebrate drift in streams. *Ecology* 46, 327-34.

In: Steer, M.W. (ed.) 1991 *Irish Rivers : Biology and Management*, pp 59-81. Royal Irish Academy, Dublin.

THE EFFECTS OF CATASTROPHIC FLOODING ON THE BENTHOS AND FISH OF A TRIBUTARY OF THE RIVER ARAGLIN, CO. CORK

Helena Twomey and Paul S. Giller

Department of Zoology, University College Cork

ABSTRACT

Benthic invertebrate density in a stream was reduced to 5% of previous levels following catastrophic flooding in August 1986. Taxon richness was reduced to 50% of pre-flood levels. By January 1987, 170 days after the flood, invertebrate density had recovered to only 35% of normal levels, while taxon richness had recovered to 75%. Diversity showed little change following the flood, but changes were observed in rank/abundance distributions, where a very truncated geometric series distribution arose after the flood. Filterer, deposit collector and predator numbers were most reduced by the flood, but filterers recovered to normal densities for a brief period six months after the disturbance. Fish size-at-age, condition and overall feeding level showed no significant reduction following the flood, nor were any changes in diet found. However, there was a short-term reduction in trout feeding level, a slight shift in trout age distribution and temporary disappearance of salmon.

INTRODUCTION

By far the most frequent agent of stream disturbance is flooding (Fisher *et al.* 1982), but the study of the effects of such disturbance on freshwater communities is still in its infancy (Sousa 1984). Periods of high water (spates) may adversely affect both invertebrate and fish populations, but the effect is very much dependent upon the intensity and frequency of spates, and rates of recovery of the biota vary widely (Fisher 1983). Severe, catastrophic flooding, such as the Teton Dam disaster in Idaho (which was followed by drying up of the stream channel), may completely denude a stream of its fish and invertebrate fauna (Minshall *et al.* 1983), but floods of such intensity rarely occur.

Catastrophic reduction in invertebrate populations following spates have been recorded by a number of authors (e.g. Moffett 1936; Allen 1951; Jones 1951; Fisher *et al.* 1982). In addition to direct mortality, benthic invertebrate populations may be adversely affected by the removal of allochthonous food (Anderson and Lehmkuhl 1968). Salmonid populations may also suffer through the destruction of eggs and fry (Elwood and Waters 1969; Seegrist and Gard 1972) which may in turn result in the virtual elimination of entire year-classes. In contrast, the effects of floods on adult salmonids, though sometimes adverse, do not seem to be as pronounced nor as predictable (Seegrist and Gard 1972). However, loss of habitat and reduced invertebrate food may result in delayed mortality, reduced growth and lower production (Allen 1951; Elwood and Waters 1969).

Flood-damaged streams regain normal conditions in time-spans ranging from a few weeks to several years, largely related to the severity and areal extent of the flood (Hynes 1970; Fisher 1983). Fisher *et al.* (1982) found that within 2-3 weeks the biota of a desert stream had recovered from a flash flood which had virtually eliminated algae and reduced invertebrate standing crop by 98%. By contrast, the macroinvertebrate community of the Teton River required over a year for recovery following flooding (Minshall *et al.* 1983). Fish populations, owing to their longer life-span, may require several years for recovery; Hanson and Waters (1974) reported a four to five year recovery period for the brook trout population in Valley Creek, Minnesota.

Unusually heavy rain in August 1986 resulted in two severe flooding events in a tributary of the River Araglin, County Cork. On 5 August, there was rainfall which would not be expected more than once in fifty years (Meteorological Service 1986); 48.4mm of rain fell in 24 hours at Moore Park, three miles from the study site, where average monthly rainfall for August is 75mm. (Rainfall data provided by Meteorological Service.) This led to a greater than 1.5 times bankful discharge at the study site. On 12 August the flood waters had receded. All mosses and macrophytes had been removed by the force of the water at the study site. Sand and silt had been largely swept away, leaving clean gravel, cobble and boulder; boulders had been moved; and the stream channel had been deepened slightly. Further heavy rain fell on 25 August (associated with Hurricane Charley), 43.2mm being recorded at Moore Park in 24 hours, again leading to greater than 1.5 times bankful discharge. As this site had been the subject of intensive study for 18 months preceding the floods, an opportunity was provided to assess the effects of such a natural disturbance event on benthic invertebrate community structure, and condition and diet of the resident fish population. Previous studies on the effects of flooding have concentrated on standing crop and production of invertebrates and fish. The present paper assesses the effects on invertebrate densities, taxon richness and diversity, rank/abundance distribution and functional group organisation for six months following the flood. The effects on trout and salmon size, condition, feeding levels and diet are also evaluated.

MATERIALS AND METHODS

The study was carried out on a 100 m stretch of a tributary close to its confluence with the River Araglin at Glenfinish Bridge (Ordnance Survey of Ireland grid reference R 877 020) in the Munster Blackwater system. Under normal flow conditions, mean channel width was 4 m, mean depth 25 cm and mean flow rate 0.8 m/second. The stream bed consisted mainly of gravel, pebble and large cobbles, with sand in small pools. The channel was well shaded by deciduous trees.

Benthic invertebrates and salmonid size, condition and diet had been regularly monitored from March 1985, 18 months prior to the flood, and monitoring has continued regularly since then. Sampling was carried out at one- (summer) or two- (winter) monthly intervals. Complete and statistically rigorous analysis of results is presented on data prior to and for the first six months following the flood. Results of the more recent monitoring are published elsewhere (Giller *et al.* in press), but the general trends indicated by these results are presented where appropriate.

Invertebrate sampling

Invertebrates were sampled using a modified (Doeg and Lake 1981) Surber sampler (area 0.13 m², 500 μm mesh). Five randomly located replicate samples were taken on each sampling occasion. The substrate was disturbed to a depth of about 5 cm for approximately 2 minutes. Material collected was preserved in the field in 70% alcohol. In the laboratory, animals were hand-sorted, identified and counted. Taxa were identified to operational taxonomic units (cf. Collier and Winterbourn 1987). In most cases this was to species, in some to order or class (e.g. Oligochaeta), and in others to family (e.g. Simuliidae).

Community variables and functional organisation

In describing invertebrate communities, several approaches have been employed in the past, including the basic community variables of mean density of invertebrates, species richness and species diversity. Patterns in the distribution of individuals amongst species have been explored through the construction of rank/abundance curves, which can be fitted to theoretical models (reviewed by Whittaker 1975; Townsend *et al.* 1983; Giller 1984; Gray 1987). A further approach has been the analysis of functional feeding group abundance and diversity (Cummins 1973; 1974; 1975; Cummins and Klug 1979; Hawkins and Sedell 1981; Townsend *et al.* 1983; Dudgeon 1984).

In an attempt to describe benthic community structure as fully as possible, all of these approaches were incorporated in the present work. On each sampling occasion, mean density of invertebrates (m^{-2}), species richness (total number of taxa) and taxon diversity ($1/\Sigma P_i^2$ where P_i is the proportion of total individuals in the ith taxon (Simpson 1949)) were calculated. Rank/ abundance curves were constructed by plotting an index of abundance (the percentage of total individuals in the community which are of the ith taxon) on a logarithmic scale against rank. These curves were compared with a geometric distribution derived from Whittaker's (1975) formula, i.e. $N_i = Nk(1-k)^{i-1}$ where N is the sum of importance values for all species in the sample, N_i the importance value for species i in the sequence from the most important to the least important species, and k is the fraction of niche space pre-empted by the most important species. In this case, k is taken as the proportion of total individuals in the community which are of the most abundant species (cf. Townsend *et al.* 1983).

An analysis of invertebrates of trophic structure was carried out. Each taxon was assigned to one of five functional groups based on the scheme of Cummins (1973), i.e. grazer/scrapers, shredders, filter feeders, deposit collectors and predators. Information on feeding biology was obtained from the literature, supported by gut content analysis in some cases (Emily Bridcot, personal communication). Where ontogenetic changes are known to occur, taxa were assigned in accordance with the dominant mode of feeding overall. The functional group classification used is given in Appendix 1. Because of the difficulty in identifying and assigning functional roles to most Chironomidae, they were excluded from this analysis.

Drift sampling

Drift samples were taken on four sampling occasions, in June, July, September and October 1986, using Cushing-Mundie samplers (500 μm mesh, Cushing 1964). Three drift samplers were placed across the stream a few metres upstream of a pool for 24 hours prior to collecting the benthos and fish samples.

Fish sampling

Fish were captured using battery-operated 400V pulsed D.C. electrofishing equipment, and held in perforated bins placed in the stream. Eels (*Anguilla anguilla* L.) were counted. Brown trout (*Salmo trutta* L.) and Atlantic salmon (*Salmo salar* L.) were anaesthetised using quinaldine (2-methylquinoline) at a concentration of approximately 15 mg/l (Gibson 1966). Each fish was then stomach-flushed (Twomey and Giller 1990). The fish was held ventral side up with its head inclined downwards over a plastic funnel leading to a 150 ml plastic storage bottle. A polyethylene tube of appropriate diameter in relation to fish size (2-7 mm external diameter) was inserted into the stomach as far as possible. A manually operated Whale gusher galley pump attached to the tubing pumped fresh water into the stomach until the stomach contents had been backflushed into the funnel. The food items washed into the container were preserved with absolute alcohol. Fish fork length (mm) and weight (on triple beam balance to nearest g) were measured to allow estimation of fish age and condition. The adipose fin was clipped to enable recaptured fish to be identified as such. Fish were returned to the perforated bins to recover, and were then released at the upstream end of the stretch in which they were caught. This sampling procedure has been shown to have no long-term effect on fish condition or feeding level (O'Farrell and McCarthy 1983; Twomey and Giller 1990).

Gut contents were identified and counted in the laboratory. The measure of feeding used (feeding level) was the number of prey items in the stomach. Fish age was determined using the length frequency distribution method (Bagenal and Tesch 1968) validated by scale reading. Condition factor was calculated as (weight(g)x100)/length(cm)^3 (Frost and Brown 1967).

Statistical analysis

Statistical analysis was carried out using MINITAB (Ryan *et al.* 1982) and SPSSX (SPSSX Inc. 1983). The data were collected over two years, with the floods occurring approximately half-way through the second year of the sampling programme. The effects of the floods may be dissociated from the effects of normal seasonal variation by using data from the first year as a control, providing no differences are found between months prior to the flood in 1986 and the same months in 1985 (M. Cole, U.C.C. Statistics Dept., pers. comm.). Therefore, data from March-July 1986 (before flooding) and from August 1986 to January 1987 (after flooding) were compared with data from the same months in the previous year. Paired t-tests were carried out to determine whether benthic invertebrate density, taxon richness, diversity and functional group organisation were significantly different from the previous year, before and after the flood. Paired t-tests were also used to test for differences within age class in salmonid length, condition and feeding level. Prior to analysis, log transformations were carried out on fish length (log x) and number of prey items per fish (log(x+1)) to normalise the data. Statistical analyses were carried out on data from 1+ and 2+ trout and 1+ salmon, as only these cohorts provided a sufficient sample size. Within cohorts, overall diet similarity between sampling occasions was assessed using Spearman rank correlation coefficients (Siegel 1956; Fritz 1974). Cluster analysis was then carried out based on the correlation coefficients, using SPSSX CLUSTER. The use of cluster analysis in dietary comparison has been demonstrated by Bortone *et al.* (1981). Wilcoxon matched-pairs tests were used to test for differences from the previous year, before and after the floods, in the relative abundance of the five most common prey taxa in salmonid diets, each of which formed over 10% of the total diet of either trout or salmon in the stream, and which together formed over 70% of trout diet and over 50% of salmon diet.

RESULTS

Benthic community structure

Benthic invertebrate density (Fig. 1(a)) was reduced from 9,383 m^{-2} in July 1986 to 494 m^{-2} in August, a reduction of 95%, and was reduced further to 361 m^{-2} in September, following the second flooding event. By January 1987 invertebrate numbers had recovered to 1,714 m^{-2}, compared with 4,907 m^{-2} in January 1986. Results from samples nearly two years after the

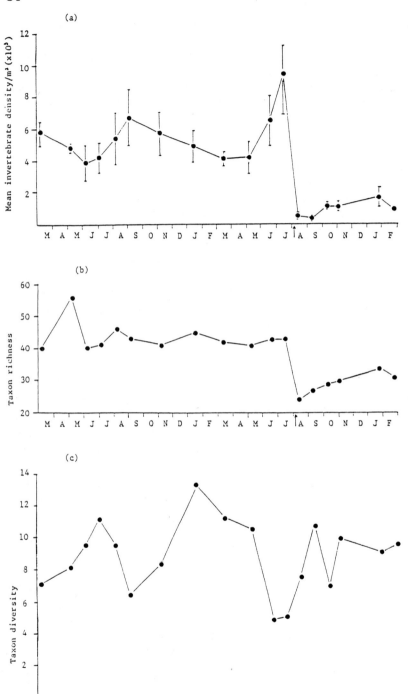

Fig. 1. Invertebrate community variables in the Araglin over a two-year period. (a) Mean invertebrate density +/- 1 standard error. (b) Taxon richness. (c) Taxon diversity. Arrow indicates timing of flooding event.

flood (May 1988) indicate that benthic invertebrate density was still below 50% of average pre-flood levels (Giller *et al.* 1990). Benthic invertebrate densities from August 1986 to January 1987 were significantly lower than from August 1985 to January 1986 (p=0.0052), whereas there was no significant difference between densities from March to July in 1985 and 1986 (p=0.43). The most affected taxa immediately following the flood included *Baetis rhodani*, which was reduced to 1.5%, Chironomidae, which were reduced to 1%, and Simuliidae, which were reduced to 0.7% of the previous month's densities. Some taxa were less severely affected; for example, *Gammarus duebeni* was reduced to 25%, and cased Trichoptera to 32%, of July densities, while *Glossosoma* sp. actually increased from 1.5 m^{-2} in July to 94 m^{-2} in August. Several taxa were not recorded in the months following the flood, including Hydrachnellae, *Amphinemura sulcicollis*, *Caenis horaria*, *Isoperla grammatica*, adult *Esolus parallelipedus*, adult *Hydraena* sp., Ceratopogonidae and *Pericoma* sp. Consequently, taxon richness (Fig. 1(b)) was reduced almost by half, from 43 taxa in July 1986 to 24 in August. Hydrachnellae and *Hydraena* sp. returned in samples by September 1986, *Isoperla grammatica* by October, Ceratopogonidae by November, *Amphinemura sulcicollis* by January 1987 and adult *Esolus parallelipedus* by February, but *Caenis horaria* and *Pericoma* sp. had not returned by February 1987 nor had some of the rarer taxa which would normally occur sporadically in samples. Thirty-four taxa were recorded in January 1987, compared with 45 taxa in January 1986.

Taxon richness from August 1986 to January 1987 was significantly lower than in the same period in the previous year (p=0.011), whereas there was no difference in the March-July comparison between the two years (p=0.68). However, while there was considerable variation, there was no significant overall difference between taxon diversity from August 1986 to January 1987 and diversity in the same period of the previous year (p=0.9).

Rank/abundance curves for each sampling occasion are plotted in Fig. 2, and the geometric distribution curves are superimposed. In 1985 (Fig. 2(a)), the curves for March and May exhibited an initially steep decline which then flattened and departed from the geometric curve. This shape, which in other months shows a tendency to become sigmoidal, suggests the data may be part of a lognormal distribution (Whittaker 1975; Townsend *et al.* 1983). The deeply concave rank/abundance curve predicted by Hughes (1986) for stable systems was not found in any month. The curve for June started to approach the geometric distribution, which was approximated in July. From August to November, the curves moved progressively towards a lognormal distribution, but the curves for January to May 1986 resembled the geometric distribution (Fig. 2(b)). The lognormal distribution returned in June and July. The curve for August 1986, immediately after the flood, was truncated and geometric in form. In September and October, the curves departed slightly from the geometric distribution, but then returned to it in November through to January and February 1987.

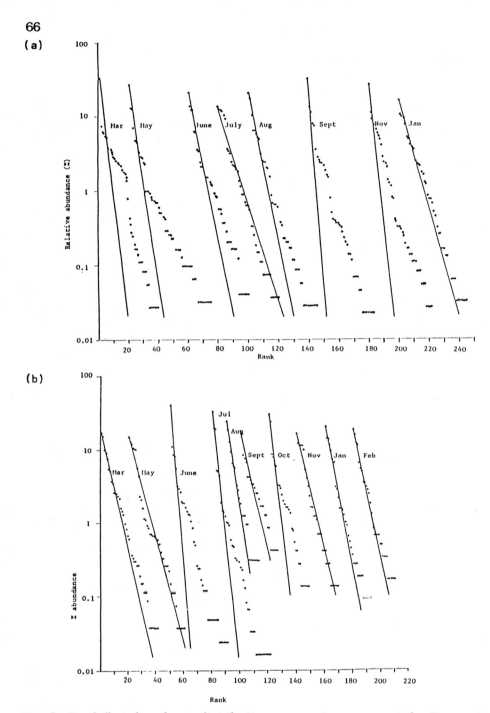

Fig. 2. Rank/log abundance distributions over a two-year period. To permit several dates to be plotted on the same axes, each distribution begins at a different point on the rank axis (cf. Whittaker 1965). Solid lines indicate hypothetical geometric series. (a) March 1985-January 1986. (b) March 1986-February 1987.

Functional group organisation

There were no significant differences in numbers or taxon richness of the five functional groups between March and July 1985 and 1986 (p=0.08-1.0). However, total numbers in all functional groups were dramatically reduced by the flood (Fig. 3(a)). Predators were not found at all in August 1986. Filterers and detritus collectors were also severely affected, being reduced to 0.8% and 3% of July levels respectively, and grazers and shredders were reduced to 26% and 18% of July levels. Filterer numbers had recovered by January 1987, but the other functional groups had attained less than half of 1986 January numbers - 31% in the case of grazers, 34% for shredders, 10% for detritus collectors, 25% for predators. For these four groups, numbers were significantly lower in the months following the flood than in the same period of the previous year (p=0.009-0.019).

Taxon richness in all functional groups was also reduced following the flood (Fig. 3(b)). Apart from predators, which were not found in August, detritus collectors were most affected. Grazers, shredders and deposit collectors had significantly fewer taxa in the months following the flood than in the same period of the previous year (p=0.011-0.015). Although filterer, predator and shredder taxon richness had recovered by January 1987, grazer and deposit collector taxon richness had not.

Drift

Mean drift density increased from 1.08 m^{-3} in June 1986 to 5.73 m^{-3} in July 1986. Following flooding and severe reduction of the benthos in August, drift density was actually higher in September (16.86 m^{-3}) than in July, and October levels were similar to July levels (5.27 m^{-3}). Thirty taxa were recorded in June, July and September, but in October only twenty taxa were recorded. Drift composition did not show apparent change following the flood; Spearman rank correlation between July and September (0.823) was higher than the correlation between June and July (0.443) or between September and October (0.745). This suggests that drift composition actually changed less between July and September (either side of the flood) than between June and July (before the flood) or between September and October (after the flood).

Fish populations

Sampling effort was standardised as far as possible but not fully quantitative, therefore no firm conclusions with respect to size of fish populations may be drawn from the catch figures presented in Appendix 2. However, the overall figures show that, whereas the trout catch was somewhat lower in August 1986, it was normal in subsequent months. Trout of age 0+ formed a greater proportion of the trout catch from August 1986 to January 1987 than in the same months in the previous year (p=0.04). Trout of 3+ and older, which were taken in most months prior to the flood, were not found on any sampling occasion in the first six months following the flood (Appendix 2),

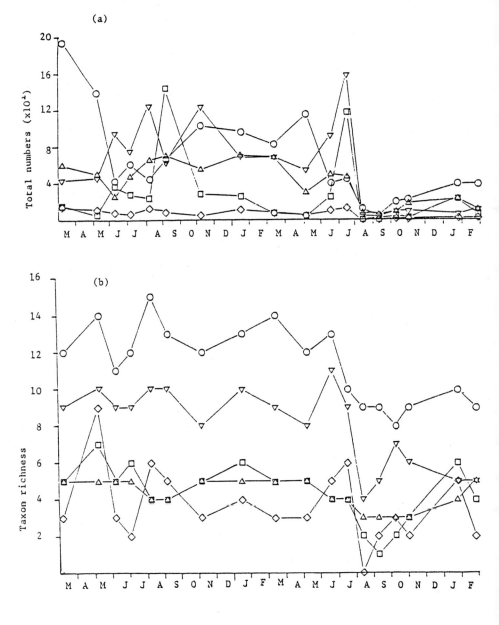

Fig. 3. Changes in functional group organisation over a two-year period in the Araglin. Grazers (O), shredders (△), filterers (□), detritus collectors (▽), predators (◇).

(a) Total numbers. (b) Taxon richness.

but had reappeared by June 1987 (pers. observation). Thus, a shift in trout population structure in favour of the younger cohorts was observed. Salmon were not caught for some months following the flood, except for one 0+ individual in August 1986 (Appendix 2). Six 0+ salmon were taken in November 1986, and good numbers of 1+ (0+ from 1986) salmon were taken in January and February, suggesting that immigration had taken place. Eels were caught in low numbers, both before and after the flood.

The proportion of recaptured (fin-clipped) salmonids in the catch (normally around 50%) was reduced following the flood (35%), and dropped further in January and February 1987 (20%), so that the percentage of recaptured salmonids from August 1986 to January 1987 was lower than for the same period of the previous year (p=0.087). Assuming that the same proportion of the population was captured on each sampling occasion, this suggests that fish population turnover in the stretch was greater than in the previous year.

Fish length, condition and feeding

Mean length and condition of 1+ and 2+ trout (Fig. 4(a,b)) showed no decrease following the flood. Feeding level (Fig. 4(c)) showed a different pattern to the previous year. Whereas in 1985 feeding level in trout remained more or less constant from July to September, feeding level in 1986 showed a decline from July to September and was lower than in the same months of the previous year. However, trout length, condition and overall feeding level from August 1986 to January 1987 were not significantly different from the same period in the previous year (p=0.12-0.91). The data thus suggest a short-term decrease in feeding level, but no effect on length or condition or on long-term feeding level. Preliminary analysis of data from continued monitoring shows no reduction in trout and salmon feeding level or condition up to May 1988, in spite of the fact that benthic invertebrate densities remained low. The length, condition and feeding level of 1+ salmon in August 1986 and January 1987 were not significantly different from the same months in the previous year (p=0.5-0.82); since no salmon were taken in September and November 1986 no comparisons could be made for these months. Salmon length was lower (p=0.030) from March to July 1986 than in the previous year, but this difference did not persist. Trout and salmon condition were lower (p=0.011-0.056) from March to July 1985 than in the following year.

Fish diets

Cluster analysis of 1+ and 2+ trout and 1+ salmon diets formed no distinct and interpretable clusters, and months following the flood were not clustered together. In several cases, fish diets in months following the flood most closely resembled diets in months before the flood. Cluster analysis of the overall diet thus suggests that the composition did not change radically following the flood. The relative abundance in the diet of five major prey items - *Gammarus duebeni, Baetis rhodani*, Chironomidae, Simuliidae, and terrestrial and aerial invertebrates - was similar before and after the flood, and there was no significant difference from the previous year, either before or after the flood (p=0.1-1.0).

DISCUSSION

Effects on benthos

Benthic invertebrate density was reduced to 5% of pre-flood levels in August, and was further reduced in September, but not all taxa were reduced to an equal extent. The most severely affected taxa (e.g. *Baetis rhodani*, Chironomidae and Simuliidae) are prominent in the drift (Twomey 1988), while taxa which are less prone to drifting (e.g. *Gammarus duebeni* and cased Trichoptera) suffered lesser reductions. By January 1987, 170 days after the flood, invertebrate density had only recovered to 35% of density for January 1986, and by May 1988, nearly two years after the flood, invertebrate density was still below 50% of average pre-flood levels (Giller *et al.* 1990). This contrasts with the faster recovery documented in previous studies described in the introduction and may be a result of the timing (summer) and areal extent of the disturbance events. Benthic invertebrate densities increased steadily but slowly from September to January following the flood, and then showed a slight reduction in February. This pattern of recovery probably represents the superposition of two contrasting processes - (a) recolonisation, and (b) the normal seasonal decline in numbers during autumn and winter.

Taxon richness was almost halved by the flood, but had recovered to 75% of normal levels by January 1987, following a linear rate of increase similar to that reported by Gore (1979) and Minshall *et al.* (1983). No further recovery had occurred up to May 1988, but taxon richness had almost fully recovered by June 1989, three years after the catastrophic flood (Giller *et al.* 1990).

Diversity showed a slight initial increase following the flood, but no overall change, in spite of the reduced number of taxa. Other workers have reported low diversity in the initial stages of colonisation (Fisher *et al.* 1982; Minshall *et al.* 1983). The slight initial increase in diversity may be explained by the differential effects of the floods upon taxa. Among the taxa which were most affected were dominant taxa in the benthos (*Baetis rhodani*, Chironomidae and Simuliidae); dominance by a few taxa accounts for the low summer values for diversity in the Araglin tributary. The reduction in relative abundance of these taxa gave rise to the observed increase in the diversity index.

The pattern of change in the shape of rank/abundance curves was complicated by the fact that the effect of the flood was superimposed on normal seasonal changes in the stream. In July 1986, the rank/abundance curve resembled the lognormal distribution, which is thought to be characteristic of complex species-rich communities (Whittaker 1975; Gray 1987). The lognormal distribution is thought to arise from the fact that the relative abundances of a large group of species are likely to be governed by the interplay of many more-or-less independent factors (May 1981). In the present case, this

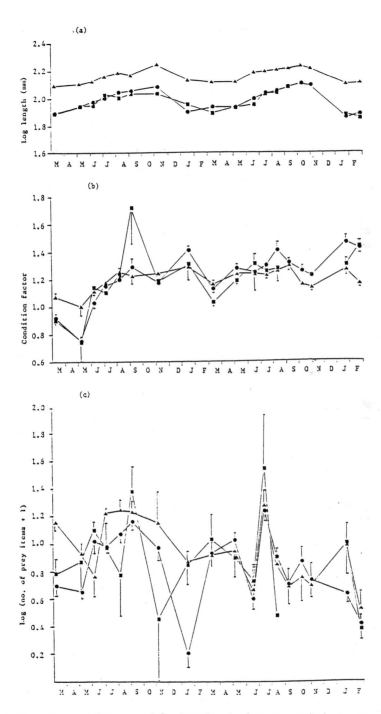

Fig. 4. Length, condition and feeding level of 1+ trout (●), 2+ trout (▲) and 1+ salmon (■) over a two-year period in the Araglin. (a) Mean length (log transformed). (b) Mean condition factor. (c) Mean feeding level (log transformed). One standard error shown where S.E. >0.02.

distribution is also characterised by high dominance by a few species. In contrast, geometric series distributions tend to arise when one major factor governs the relative abundance of different species (May 1981), which is characteristic of severe environments, of habitats following disturbance events such as pollution, and of the early stages of succession (Whittaker 1975; May 1981; Gray 1987). The curve for August 1986, immediately following the first flood, fitted the geometric distribution, which is normally displayed in the winter in the study stream, although the curve for July 1985, when numbers were still low as a result of a late spring, also fitted the geometric distribution. In all these cases, dominance was low. The dynamics model of Hughes (1986) also predicts that a distribution resembling the logseries, a statistically realistic variant of the geometric series (May 1981), may be found in samples from a disturbed community, caused by the reduction of dominant taxa. The logseries may, however, simply describe small samples from a lognormally distributed community (Preston 1948), resulting from the lower species number and density.

The gradual shift from the geometric towards a lognormal-type distribution between August and October 1986 is typical of the changes which take place in species abundance distributions during succession or recovery from a major disturbance as some taxa reinstate their numerical dominance (Whittaker 1972; May 1981; Giller 1984; Gray 1987); however, the distribution returns to the geometric between October and February 1987, possibly as a result of the normal seasonal changes which reduce dominance and give rise to the geometric distribution in winter.

Functional group organisation

Immediately after the flood, predators had totally disappeared, filterers and deposit collectors had suffered a severe reduction in numbers, and grazers and shredders were reduced to a lesser extent. Grazers predominated during the early stages of recolonisation, as was reported by Minshall *et al.* (1983). Filterers were the only group which had recovered to normal levels by January 1987, possibly owing to the year-round reproductive capacity of Simuliidae (Hynes 1970); however, normal filterer numbers in January are low and subsequent samples revealed that recovery was only short-lived (Giller *et al.* 1990). Deposit collectors showed the poorest recovery, attaining only 10% of the previous January levels, suggesting that the buildup of detritus was slow or that deposit collectors were slow recolonisers. The former seems more likely, since *Baetis rhodani*, a collector, was prevalent in the drift following the flood. Grazers and shredders attained 31% and 34% respectively of 1986 January levels; this may reflect the availability of their food sources, suggesting that algae and coarse particulate organic matter became available more rapidly than fine detritus and invertebrate prey. However, they were not reduced to the same extent as the other functional groups as a result of the flood. Predators showed a fairly good recovery within six months; there was no evidence of the long lag (about 250 days) in the establishment of predators reported by Gore (1979) and Minshall *et al.* (1983).

Taxon richness within functional groups showed similar patterns, with predator, deposit collector and filterer taxon richness being most reduced by the immediate effects of the flood. Filterer and predator taxon richness had recovered to normal by January 1987.

Drift

Drift density did not decrease following the flood. Indeed, drift density was at its highest in September, when invertebrate density was lowest, which suggests the possibility of positive (behavioural) drift to avoid unfavourable conditions. Minshall *et al.* (1983) also found that drift density was not related to benthic invertebrate density, and concluded that potential colonists (especially *Baetis* and Chironomidae) can move over relatively long distances (km) in a reasonably short time (days). Drift is one of the methods of colonisation of new substrate areas by benthic invertebrates (Townsend and Hildrew 1976; Giller and Campbell 1989). When flooding is localised leaving undisturbed upstream reaches, initial recolonisation by drift is important (Fisher *et al.* 1982). Drift was probably the main source of recolonisation in this case, as the flooding occurred later than the main reproductive period for stream invertebrates in northern temperate streams (Hynes 1970). Upstream migration seems unlikely, as the area downstream of the study site was probably also badly affected by the flood.

Fish populations

Trout populations do not appear to have been adversely affected by the flood; the reduction observed on August 1986 may have resulted from temporary physical displacement as described by Elwood and Waters (1969). In particular, 0+ trout, which are supposedly more vulnerable to flood damage than older fish (Elwood and Waters 1969; Hynes 1970; Seegrist and Gard 1972), were actually more abundant than in the previous year. Salmon appear to have suffered at least temporary displacement from the stretch, although the salmon catch had recovered well by January. It has been shown that intra-specific competition from trout can restrict salmon to less favoured habitat (Kennedy and Strange 1986); it is possible that under the unfavourable conditions following the flood, salmon may have been excluded from the stretch entirely. In addition, the reduction in benthic invertebrate densities could have more severe effects on salmon than on trout feeding, since terrestrial and aerial invertebrates form a large part of the normal trout diet in the study stream (Twomey 1988). The recovery of 1+ salmon numbers in January and February suggests that migration rather than mortality may have been responsible for the decline in the number of salmon caught. The observed reduction in the proportion of recaptured salmonids over the six months following the flood also suggests increased fish movement and population turnover.

Fish length, condition, feeding level and diet

Although trout did show a temporary reduction in feeding level as a result of the flood, condition and size-at-age appear to have been unaffected, and feeding level recovered by November. It seems surprising that fish should not have suffered more as a result of the dramatic reduction in benthos, as other authors have found reduced growth and lower production in salmonid populations following floods (Allen 1951; Elwood and Waters 1969). One explanation may lie in the feeding habits of the fish in this stream system. Drift feeding and benthic feeding are both of importance to salmonids in the Araglin tributary (Twomey 1988). Since drift density remained high following the flood, the availability of drifting food would not have been reduced in the same way as the availability of benthic food. Also terrestrial food was important in salmonid, particularly trout, diets in the Araglin (Twomey 1988). This food source was unaffected by the flood. Thus, the continuing availability of aquatic, and particularly terrestrial, invertebrates in the drift may have allowed salmonids to obtain their normal food ration, in spite of the catastrophic reduction in benthic invertebrates. However, no change was found in overall diet composition following the flood, and the relative abundance of terrestrial invertebrates in the diet, although tending to be higher, was not significantly greater than in the previous year. The maintenance of drift density despite catastrophic reduction in the benthos leaves another unanswered problem.

An alternative explanation for the lack of effect of reduction in invertebrate density on fish may be proposed. If fish populations in the Araglin tributary are limited by factors other than food supply, for example by territory size, spawning or nursery areas, or availability of suitable foraging sites, food might normally be superabundant to the extent that even catastrophic reduction in the benthos would have little effect on fish feeding. Under these circumstances, fish would normally consume only a small fraction of available prey, as was suggested by Allan (1982). This would make it unlikely that predatory fish play a large role in structuring the benthic invertebrate community in this system.

REFERENCES

ALLAN, J.D. 1982 The effects of reduction in trout density on the invertebrate community of a mountain stream. *Ecology* 63, 1444-5.

ALLEN, K.R. 1951 The Horokiwi stream. A study of a trout population. *Bulletin of the Marine Department of New Zealand Fisheries* 10, 1-238.

ANDERSON, N.H. and LEHMKUHL, D.H. 1968 Catastrophic drift of insects in a woodland stream. *Ecology* 49, 198-205.

BAGENAL, T.B. and TESCH, F.W. 1968 Age and growth. In T.A. Bagenal (ed.), *Methods for assessment of fish production in fresh waters*, 101-36. IBP Handbooks no. 3. Oxford. Blackwell.

BORTONE, S.A., SEIGEL, D. and OGLESBY, J.L. 1981 The use of cluster analysis in comparing multisource feeding studies. *Northeast. Gulf Sci.* 5, 81-6.

COLLIER, K.J. and WINTERBOURN, M.J. 1987 Faunal and chemical dynamics of some acid and alkaline New Zealand streams. *Freshwater Biology* 18, 227-40.

CUMMINS, K.W. 1973 Trophic relations of aquatic insects. *Annual Review of Entomology* 18, 183-206.

CUMMINS, K.W. 1974 Structure and function of stream ecosystems. *Bioscience* 24, 631-41.

CUMMINS, K.W. 1975 Macroinvertebrates. In B.A. Whitton (ed.), *River ecology*, 170-98. Oxford. Blackwell.

CUMMINS, K.W. and KLUG, M.J. 1979 Feeding ecology of stream invertebrates. *Annual Review of Ecology and Systematics* 10, 147-72.

CUSHING, C.E. 1964 An apparatus for sampling drifting organisms in streams. *Journal of Wildlife Management* 28, 592-4.

DOEG, T. and LAKE, P.S. 1981 A technique for assessing the composition and density of the macroinvertebrate fauna of large stones in streams. *Hydrobiologia* 80, 3-6.

DUDGEON, D. 1984 Longitudinal and temporal changes in functional organization of macroinvertebrate communities in the Lam Tsuen River, Hong Kong. *Hydrobiologia* 111, 207-17.

ELWOOD, J.W. and WATERS, T.F. 1969 Effects of floods on food consumption of a stream brook population. *Transactions of the American Fisheries Society* 98, 253-62.

FISHER, S.G. 1983 Succession in streams. In J.R. Barnes and G.W. Minshall (eds), *Stream ecology. Application and testing of general ecological theory* 7 -27. New York. Plenum Press.

FISHER, S., GRAY, L., GRIMM, N. and BUSCH, D. 1982 Temporal succession in a desert stream ecosystem following flash flooding. *Ecological Monographs* 52, 93-110.

FRITZ, E.S. 1974 Total diet comparison in fisheries by Spearman Rank Correlation Coefficients. *Copeia* 1, 210-14.

FROST, W.E. and BROWN, M.E. 1967 *The trout*. London. Collins.

GIBSON, R.N. 1966 The use of the anaesthetic quinaldine in fish ecology. *Journal of Animal Ecology* 36, 295-301.

GILLER, P.S. 1984 *Community structure and the niche*. London. Chapman and Hall.

GILLER, P.S. and CAMPBELL, R.N.B. 1989 Colonisation patterns of mayfly nymphs (Ephemeroptera) on implanted substrate trays of different size. *Hydrobiologia* 178, 59-71.

GILLER, P.S., SANGPRADUB, N. and TWOMEY, H. (in press) Catastrophic flooding and macroinvertebrate community structure. *Verhandlungen Internationale Vereinigung für Theoretische und Angewandte Limnologie* 24.

GORE, J.A. 1979 Patterns of initial benthic recolonization of a reclaimed coal strip-mined river channel. *Canadian Journal of Zoology* 57, 2429-39.

GRAY, J.S. 1987 Species-abundance patterns. In J.H.R. Gee and P.S. Giller (eds), *Organization of communities, past and present*, 53-67. Oxford. Blackwell.

HANSON, D.L. and WATERS, T.F. 1974 Recovery of standing crop and production rate of a brook trout population in a flood-damaged stream. *Transactions of the American Fisheries Society* 103, 431-9.

HAWKINS, C.P. and SEDELL, J.R. 1981 Longitudinal and seasonal changes in functional organization of macroinvertebrate communities in four Oregon streams. *Ecology* 62, 387-97.

HUGHES, R.G. 1986 Theories and models of species abundance. *The American Naturalist* 128, 879-99.

HYNES, H.B.N. 1970 *The ecology of running waters.* Liverpool University Press.

JONES, J.R.E. 1951 An ecological study of the River Towy. *Journal of Animal Ecology* 20, 68-86.

KENNEDY, G.J.A. and STRANGE, C.D. 1986 The effects of intra- and inter-specific competition on the distribution of stocked juvenile Atlantic salmon, *Salmo salar* L., in relation to depth and gradient in an upland trout, *Salmo trutta* L., stream. *Journal of Fish Biology* 29, 199-214.

MAY, R.M. 1981 Patterns in multi-species communities. In R.M. May (ed.), *Theoretical ecology : principles and applications*, 197-227. Oxford. Blackwell.

METEOROLOGICAL SERVICE 1986 *Monthly weather bulletin : Supplement to August 1986.* Dublin. Meteorological Service.

MINSHALL, G.W., ANDREWS, D.A. and MANUEL-FALER, C.Y. 1983 Application of island biogeographic theory to streams: macroinvertebrate recolonization of the Teton River, Idaho. In J.R. Barnes and G.W. Minshall (eds), *Stream ecology. Application and testing of general ecological theory*, 279-97. New York. Plenum Press.

MOFFETT, J.W. 1936 A quantitative study of the bottom fauna in some Utah streams variously affected by erosion. *Bull. Univ. Utah biol. Ser.* 26, Part 9.

O'FARRELL, M.M. and McCARTHY, T.K. 1983 An evaluation of stomach flushing as a fishery technique. *Proceedings of the Third British Freshwater Fish Conference*, 59-70. University of Liverpool.

PRESTON, F.W. 1948 The commonness and rarity of species. *Ecology* 29, 254-83.

RYAN, T.A., JOINER, B.L. and RYAN, B.F. 1982 *Minitab reference manual.* Pennsylvania. Minitab Incorporated.

SEEGRIST, D.W. and GARD, R. 1972 Effects of floods on trout in Sagenhen Creek, California. *Transactions of the American Fisheries Society* 101, 478-82.

SIEGEL, S. 1956 *Nonparametric statistics for the behavioral sciences.* Tokyo. McGraw-Hill Kogakusha.

SIMPSON, E.H. 1949 Measurement of diversity. *Nature* 163, 688.

SOUSA, W.P. 1984 The role of disturbance in natural communities. *Annual Review of Ecology and Systematics* 15, 353-91.

SPSSX INC. 1983 *SPSSX user's guide.* New York. McGraw-Hill.

TOWNSEND, C.R. and HILDREW, A.G. 1976 Field experiments on the drifting, colonization and continuous redistribution of stream benthos. *Journal of Animal Ecology* 45, 759-72.

TOWNSEND, C.R., HILDREW, A.G. and FRANCIS, J. 1983 Community structure in some southern English streams : the influence of physico-chemical factors. *Freshwater Biology* 13, 521-44.

TWOMEY, H. 1988 Fish predation in relation to lotic invertebrate community structure. Unpublished Ph.D. thesis, National University of Ireland.

TWOMEY, H. and GILLER, P.S. 1990 Stomach flushing and individual Pan-jet tattooing of salmonids : an evaluation of the long-term effects on two wild populations. *Aquaculture and Fisheries Management* 21, 137-42.

WHITTAKER, R. 1965 Dominance and diversity in land plant communities. *Science* 147, 250-60.

WHITTAKER, R.H. 1972 Evolution and measurement of species diversity. *Taxon* 21, 213-51.

WHITTAKER, R.H. 1975 *Communities and ecosystems* (2nd edition). New York. Macmillan.

APPENDIX 1

Species list and functional feeding group classification

G: Grazers/scrapers
S: Shredders
F: Filterers
D: Deposit collectors
P: Predators

Nematoda	D
Nematomorpha	
Tricladida	
Polycelis nigra (Muller) / *tenuis* (Ijima)	D
Hirudinea	
Batracobdella paludosa (carena)	P
Erpobdella octoculata (L.)	P
Piscicola geometra (L.)	P
Oligochaeta	D
Mollusca	
Ancylus fluviatilis Muller	G
Aplexa hypnorum (L.)	G
Lymnaea peregra (Muller)	G
Pisidium spp	G
Potamopyrgus jenkinsi (Smith)	G
Hydracarina	P
Malacostraca	
Asellus meridianus Racovitza	S
Gammarus duebeni Lilljeborg	S
Ostracoda	F
Odonata	P
Ephemeroptera	
Baetis muticus (L.)	D
Baetis rhodani (Pict.)	D
Caenis horaria (L.)	D
Ecdyonurus dispar (Curtis)	G

Ephemerella ignita (Poda)	D
Heptagenia sulphurea (Muller)	D
Paraleptophlebia submarginata (Stephens)	D
Rhithrogena semicolorata (Curtis) / *haarupi* (Est.-Pet.)	G

Plecoptera

Amphinemura sulcicollis (Stephens)	S
Brachyptera risi (Morton)	G
Chloroperla torrentium (Pictet)	S
Isoperla grammatica (Poda)	P
Leuctra hippopus (Kempny)	S
Leuctra inermis (Kempny)	S
Nemoura cinerea (Retzius)	S
Nemurella picteti (Klapalek)	S
Protonemura meyeri (Pictet)	S

Coleoptera

Agabus spp	P
Brychius elevatus (Panzer)	G
Coelostoma spp	D
Elmis aenea (Muller)	G
Esolus parallelipedus (Muller)	G
Haliplus spp	G
Hydraena spp	D
Hydrophilidae	D
Hydroporus spp	P
Ilybius spp	P
Limnius volckmari (Panzer)	G
Orectochilus spp	P
Oreodytes spp	P
Oulimnius tuberculatus (Muller)	G
Potamonectes spp	P
Stictonectes lepidus	P

Diptera

Ceratopogonidae	D
Chironomidae	
Dixidae	D
Pericoma spp	D
Simuliidae	F
Tipulidae	

Hemiptera

Velia spp	P
Corixidae	P

Trichoptera

Agapetus fuscipes Curtis	G
Glossosoma spp	G
Hydropsyche angustipennis (Curtis)	F
Hydropsyche fulvipes (Curtis)	F
Hydropsyche instabilis (Curtis)	F
Hydropsyche pellucidula (Curtis)	F
Hydropsyche siltalai Dohler	F
Philopotamus montanus (Donovan)	F
Plectrocnemia conspersa (Curtis)	P
Plectrocnemia geniculata McLachan	P
Polycentropus kingi McLachan	P
Rhyacophila dorsalis (Curtis)	P
Cased caddis	

APPENDIX 2

Araglin: Numbers of fish caught on each sampling occasion

	Trout					Salmon			Eels
	0+	1+	2+	3+	4+	0+	1+	2+	
March 1985		12	11	2	1		7	6	
May **		2	5				4	1	
June		15	8	5			8		
July	3	35	14	1		1	4	1	4
August	3	28	11	3			5		3
September	6	17	10	1			3	1	
November	8	24	6			2	2		3
January 1986		6	19	4	1		3	1	2
March		22	13	2			3		
May		19	13	2			7		1
June		24	6	1			5		1
July	6	26	7				3		5
August	4	13	2				1		3
September	10	13	6						2
October	23	17	3						
November	14	9	6			6			1
January 1987		9	6				7	3	
February		30	9				17		2

**Fishing curtailed owing to problems with electrofishing apparatus.

Part 3

Management of biological components in rivers

In: Steer, M.W. (ed.) 1991 *Irish Rivers : Biology and Management*, pp 85-98. Royal Irish Academy, Dublin.

AQUATIC PLANT MANAGEMENT IN IRISH RIVERS

Joseph M. Caffrey
Central Fisheries Board, Mobhi Boreen, Glasnevin, Dublin
and
Botany Department, University College Dublin

ABSTRACT

In many Irish rivers aquatic plants present serious flood hazards in addition to adversely affecting the amenity use of these water-courses. These problems have been exacerbated as a result of man's interference with the natural habitat, mainly through cultural eutrophication and drainage. Weed control strategies embrace four broad categories - mechanical, chemical, environmental and biological - of which mechanical and chemical are the most widely used. The results of trials, using manual cutting and the aquatic herbicides Casoron and Midstream, are described. These results reveal that manual cutting offers only short-term control and may actually stimulate growth. As a consequence two cuts in a single growing season may be required. More satisfactory levels of control were achieved when chemical treatments were targeted against specific weed problems, although the degree of success obtained was often species-specific. The risk to fisheries or potable water supplies from the correct use of aquatic herbicides is minimal and is safeguarded by formal clearance procedures.

INTRODUCTION

Ireland contains approximately 13,500 km of main river and stream channel (Anon. 1986). These lotic systems range in character from small torrential upland streams to large slow-flowing lowland rivers. Depending upon the perspective of the user, rivers serve a diversity of purposes. To the drainage engineer their primary role is the speedy transport of water from the land

85

to the sea. Their role as seen by the municipal engineer is to supply an adequate volume of clean water which contains the minimum amount of filter-clogging vegetation or debris. To the angler rivers provide a diversity of habitats which provide cover and feeding areas for both coarse and game fish. To the unthinking or uncaring in our community rivers serve as convenient receiving systems for domestic, agricultural, municipal or industrial effluents. The nature and abundance of the aquatic plant communities present in these rivers will determine, to a large extent, the level of success achieved by these various user groups.

Aquatic plants play an integral role in the structure and functioning of riverine ecosystems. However, people differ widely in their attitudes to aquatic plants. Some view them as mere obstacles to the free flow of water while others see them as providing inchannel conditions which favour the establishment of diverse biotic communities. This diversity of opinion stems from a lack of understanding of the role which aquatic plants play in the lotic habitat.

Aquatic plants in a naturally balanced river system serve a number of important functions. Among these are included:

- the provision of substrates for the establishment of periphyton and macro-invertebrates;
- the provision of a direct and indirect food source for macro-invertebrates, fish and wildfowl;
- the provision of shelter for young and fodder fish, and concealment of predatory fish;
- the provision of spawning substrates for coarse fish species;
- the consolidation of river beds or banksides;
- the filtration of suspended solids and available nutrients (nutrient stripping) from the water;
- the production of oxygen as a by-product of the photosynthetic process;
- the provision of an aesthetically pleasing watercourse.

In Ireland, the length of clean undisturbed river channel which supports a balanced aquatic flora has decreased sharply in the last two decades. This has resulted primarily from the progressive spread of cultural eutrophication in these channels (An Foras Forbartha 1986). The impact of arterial drainage on rivers in Ireland has also seriously affected the natural floral balance. In many instances these effects on the lotic habitat are reflected by significant increases in maximum macrophytic biomass and by changes from diverse plant communities to those dominated by one or two opportunistic species. This can adversely affect the beneficial use of rivers in a number of ways. It may lead to:

- an increased risk of flooding;
- excessive siltation in the channel;
- impairment of recreational resource facilities such as boating, angling, swimming;
- fish mortality resulting from large diurnal fluctuations in oxygen concentrations and/or oxygen depletion on the death and decay of the plant material;

- clogging of filters at water treatment, or other, installations.

This paper outlines a range of aquatic plant control strategies which are available for use on rivers. It also presents some preliminary experimental results achieved under Irish conditions using some of these control procedures. The merits and demerits of these techniques are briefly discussed.

AQUATIC PLANT CONTROL STRATEGIES

To achieve the level of performance required of a river by the various user groups it is generally necessary to manage or control the inchannel vegetation. The plant species or groups which, in Irish rivers, are responsible for most of the weed-related problems are presented in Table 1. Presented also are those stimuli which may result in the accelerated growth recorded, the range of river types affected, the appropriate control strategies and the likelihood of achieving an acceptable level of control.

The weed control procedures most widely adopted internationally embrace four broad categories. These are briefly described below.

Mechanical control

Traditionally aquatic vegetation was removed from channels manually, using scythes, slashers and rakes. Because this method is costly in terms of labour, it has become restricted to small specialised watercourses. In these waters this method is very effective because it allows the vegetation to be 'tailored' to meet the specific needs of the user.

Between 1950 and 1970 major advances in engineering technology produced a wide range of elaborate weed-cutting and harvesting machines for this market. These include specialised weed-cutting boats, hydraulically operated weed-cutting buckets and tractor-mounted flail mowers, among others (Arts and van Wijk 1978). This machinery allows long sections of overgrown river channel to be effectively cut with the minimum of labour costs. The capital cost of purchasing this machinery, however, is high and, as with manual control, vegetative regrowth is often rapid. As a consequence, two or even three cuts during a single season may be required.

Environmental control

Bankside shading is the most commonly adopted environmental technique used for the control of aquatic plants in rivers. The reduction of light has long been considered a technique for the reduction of excess growths of aquatic plants (Dawson 1981). Studies in the U.K. have shown that the maximum seasonal biomass of several different aquatic plant species in streams was reduced when the availability of light to the water surface was restricted (Dawson and Kern-Hansen 1979). Few quantitative data on the efficiency of this technique for aquatic plant control under Irish conditions have been collected.

Biological control

Biological control involves the use of living organisms to control nuisance weed growths. For the control of aquatic plants the use of grass carp has been widely recommended (Seagrave 1988). The main research effort on this subject has been undertaken in England and the Netherlands (Stott 1977; van Zon 1977). Results to date indicate that in any stocking programme careful consideration must be given to stock density and size of fish introduced if the watercourse is not to be completely denuded of vegetation. As it is illegal to introduce new fish species into Irish waters the merits or demerits of this fish as a weed management tool will not be explored in this paper.

Chemical control

In the U.K. clearance for the use of any pesticide is granted under the Pesticides Safety Precautions Scheme (PSPS) which is a formally agreed scheme between the British government and the agrochemical industry (Ministry of Agriculture, Fisheries and Food 1979). The object of the scheme is to safeguard people, animals and the environment against risks arising from the use of pesticides. There is currently no legislation in Ireland governing the use of herbicides. However, a new chemical registration scheme is presently being implemented. Under this scheme any chemical being sold in Ireland must be registered with the Department of Agriculture and full details of environmental, toxicity and residue testing must be available.

Only eight herbicides are cleared under the PSPS for use in or near water. This clearance limits both maximal concentrations and minimal intervals between herbicide treatment and use of the treated water for irrigation or potable supply (Brooker and Edwards 1975). Herbicides which are used for aquatic weed control fall into two main groups. Group one is applied directly to emergent or floating vegetation as a spray. The second group is applied to the water as liquid, gel or in a granular formulation. It is within this second group that the herbicides which exert the greatest control on nuisance submerged plant growths are contained.

The aquatic herbicides which may be most appropriate for use in Irish rivers and upon which most trials work has been concentrated are Casoron and Midstream. The former is a systemic herbicide and is available in granular formulation. The active ingredient, dichlobenil, is absorbed into the plant by root uptake from the hydrosoil. In the plant the chemical acts primarily on dividing meristems. Casoron is specially formulated for use in still or slow-flowing watercourses.

Midstream is a contact herbicide containing diquat in a viscous gel formulation. It is specially formulated for the control of weeds in flowing waters. Once absorbed into the plant tissue, diquat interferes directly with the plant's photosynthetic mechanism.

Trials work with Clarosan and Roundup only began in 1988 and no conclusive data are yet available.

PRELIMINARY RESULTS

Little experimentation has been conducted on the use and efficacy of weed control strategies in Irish rivers. The results presented below are among the first quantitative accounts of this type of research under Irish conditions. As trial work was only initiated in 1987, and many of the experiments are ongoing at present, the results presented must be regarded as preliminary findings which may be reviewed in the light of further research.

The main findings are presented graphically in Fig. 1 (A to F) and are briefly described below.

Manual control

Experiment 1

In the River Shiven at Mountbellew, Co. Galway, horned pondweed (*Zannichellia palustris* L.) is the dominant plant species. By June 1986 monospecific stands of this submerged plant were seriously impeding the flow and had accumulated sufficient debris to detract from the aesthetics of the environ. In early June the standing crop was reduced from 253 to 10 g DW m^{-2} using scythes and rakes. Regrowth of this silt-loving elodeid was initially slow although in the eight weeks between early August and late September vigorous vegetative regrowth increased the standing crop from 20 to 174 DW m^{-2} (Fig. 1A). Thereafter the growth rate slowed appreciably as the plant entered the decline phase of its growth cycle.

Experiment 2

In one of the many sections of the River Liffey where dense silt-depositing stands of fennel pondweed (*Potamogeton pectinatus* L.) presented year-round inchannel problems, the effects of an early season manual cut were examined. An adjacent section of river which supported a similar standing crop of this invasive perennial was left uncut and used for comparative purposes. The manual cut reduced vegetative biomass in the cut section from 180 to 52 g DW m^{-2} (Fig. 1B). Vegetative regrowth began more or less immediately following the cut and within five weeks the standing crop had more than trebled. Between mid-April and mid-June the growth rate in the cut section exceeded that in the uncut control area and, at the end of this nine-week period, the cut channel supported a marginally greater vegetative crop than that recorded from the uncut channel (Fig. 1B). From mid-June to the end of the sampling period in November the differential in growth rates between the sections was negligible.

90

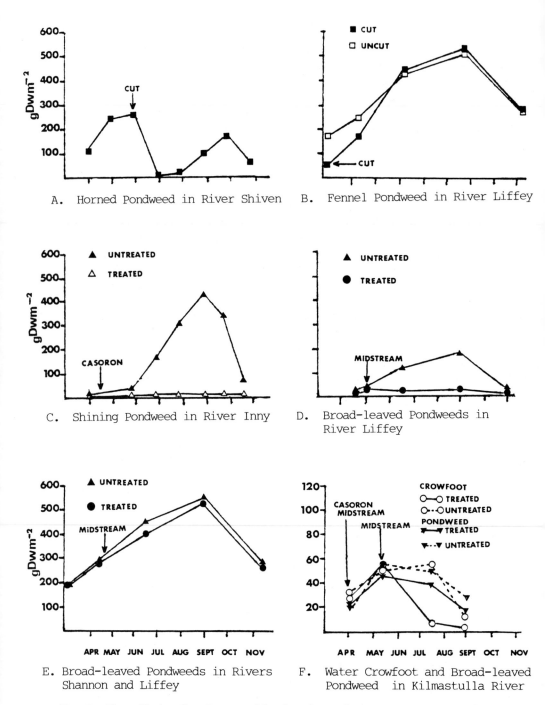

A. Horned Pondweed in River Shiven

B. Fennel Pondweed in River Liffey

C. Shining Pondweed in River Inny

D. Broad-leaved Pondweeds in River Liffey

E. Broad-leaved Pondweeds in Rivers Shannon and Liffey

F. Water Crowfoot and Broad-leaved Pondweed in Kilmastulla River

Fig. 1. The effects of cutting and herbicide application on a range of macrophytes in Irish rivers.

Chemical control

Experiment 3

Long sections of the River Inny, and many other Irish rivers, support dense monospecific stands of shining pondweed (*Potamogeton lucens* L.), a broad-leaved, submerged macrophyte. The growth cycle of this plant in the River Inny is such that virtually no vegetation emerges from the hydrosoil until the end of May. Between this and the end of September vegetative growth is rapid and a standing crop of in excess of 400 g DW m^{-2} may be produced in this short growing season. Because of the extent of obstructive inchannel vegetation during the growing season, flood hazards are presented and the channel, which previously offered an excellent water-based recreational resource, is incapable of serving any amenity interest.

As the flow velocity in this drained channel is slow (15 c s^{-1} at the time of spraying), treatment with either Casoron or Midstream should prove equally effective. However, as it is desirable to spray in April/May if ecological problems associated with fish spawning and wildfowl nesting are to be minimised, and as no leafy shoots onto which the Midstream strands can attach are present at this time, Casoron is the obvious choice. This product has a pre-emergence activity so the target plant can be suppressed before it emerges from the mud. A slow-release (G SR) granular formulation of Casoron was applied to a 1 km long stretch of river in April 1987 and to a similar length of channel in April 1988. The chemical was applied to a central channel area measuring 8 m in width. Two marginal bands, each approximately 4 m wide, remained untreated. It was hoped that aquatic plants in these marginal areas would be unaffected by the treatment and provide a refuge for aquatic insects, fish and wildfowl. An untreated control channel situated upstream of the treated area was selected for comparative purposes.

In the treated channel the standing crop of shining pondweed did not exceed 12 g DW m^{-2} during the 1987 and 1988 sampling periods (Fig. lC). Along the untreated margins of this Casoron-treated section an abundant and diverse aquatic flora grew and provided sanctuary for the resident aquatic fauna. So effective was the control in this area that a highly prestigious international match angling festival was held here in the summer of 1987. In the untreated control area, on the other hand, dense monospecific stands of shining pondweed filled the channel for most of the growing season. The pursuance of angling or any other water-based amenity activity was rendered impossible as a consequence of this weed growth.

Experiment 4

In a moderately fast-flowing (35 c s^{-1} at time of spraying) section of the River Liffey, which is typical of many kilometres of this and other highly prized salmonid fisheries, dense growths of broad-leaved pondweed species, mainly various-leaved pondweed (*Potamogeton gramineus* L.), are commonplace. Because of the channel depth (>1 m) and the inaccessibility to machinery of these channels they are difficult to manage efficiently. If it was possible to control the weed chemically, without impacting the resource or adversely affecting potable water supplies, this would provide a convenient and cost-effective management tool to those responsible for maintenance of the watercourse.

Because of flow constraints Midstream, in a "sticky" gel formulation, was the only chemical which could be effectively applied to the plants. Sprayed onto the water using a specialised applicator, the strands of chemical-laced gel rapidly sink and attach to the target plants. Localised treatment is possible using this chemical.

The section selected for study was treated in late April when sufficient submerged plant material was present for attachment of the gel. An ecologically similar control section, upstream of the treated area, was selected for comparative purposes.

Following Midstream application the treated plants died back within two weeks. A small carpet of low-growing pondweed, insufficient to interfere with angling, remained in the channel through the growing season (Fig. 1D). In the untreated channel vigorous vegetative growth continued through the season, producing a maximum standing crop of 182 g DW m^{-2} in September. Much of this consisted of surface-growing leafy stands of pondweed which present "nightmare" conditions for anglers.

Experiment 5

In view of the poor level of control which was achieved when fennel pondweed, in the River Liffey (Fig. 1B), was manually cut, a number of ecologically similar, fennel pondweed-dominated sites on the Rivers Shannon and Liffey were selected for chemical treatment. Because of the erosive flow velocities (1 m s^{-1}) which can operate at these sites Midstream was the only product which could be used. Untreated control areas were also selected for examination.

From Fig. 1E it is apparent that Midstream application had little effect upon the growth of this robust perennial. Some minor retardation of growth between May and September resulted in a marginally reduced standing crop in the treated area but this was insignificant in relation to the bulk of plant material which was present in the river. These preliminary results suggest that fennel pondweed is resistant to Midstream but further experimentation is necessary before a conclusive statement may be made.

Experiment 6

In many fast-flowing and gravelled streams and rivers, particularly those which are not subjected to chronic organic pollution, dense monospecific stands of water crowfoot (*Ranunculus* spp) often present flood hazards during the spring and summer months. These excessive growths are most prolific in channels where bankside vegetation has been removed, allowing maximum penetration of incident light to the river bed.

In sections of the River Kilmastulla, a tributary of the River Shannon, mixed stands of water crowfoot and broad-leaved pondweed (*Potamogeton natans* L.) annually overgrow the channel. To avoid flooding problems this channel is manually cut each summer by County Council operatives. This is a costly, labour-intensive and relatively ineffective operation. To investigate the possibility of improving the efficiency of weed management on this river an experiment using aquatic herbicides was initiated.

In early April separate channel sections were sprayed with Midstream and Casoron G SR. At the time of spraying the river was in flood and highly coloured. When examined six weeks later little or no difference was recorded in standing crop values from treated and control sections (Fig. 1F). This result was not unexpected because of the adverse conditions operating when the chemicals were applied. Because the normal flow velocity in this river (average 30 c s[-1)]) exceeds the threshold for effective Casoron use no further trials with this product were conducted. It was decided, however, to apply a second dose of Midstream because, at the time of the first application, it is likely that most of the chemical was inactivated by suspended solids in the flood water (Fox *et al.* 1986).

Reference to Fig. 1F shows that a high level of control with water crowfoot was achieved, even taking account of the late application date. The significantly reduced standing crop recorded from the untreated zone in September reflects the natural seasonal decline among the plants. An unsatisfactory level of control was achieved with broad-leaved pondweed, a plant which is moderately resistant to Midstream.

DISCUSSION

Aquatic plants play an integral role in the functioning of the aquatic ecosystem. They have a central position in the network of ecological relations between nutrients, plankton, macro-invertebrates, fish and avifauna (de Nie 1987). Indiscriminate aquatic plant removal may therefore serve to destabilise the aquatic ecosystems.

The ideal form of weed control is one where aquatic plant biomass is reduced to, and maintained at, an acceptable level without loss of species diversity and with minimal disturbance to the habitat (Barrett 1978). In practice it is difficult to determine what level of plant growth is acceptable and there must be a degree of flexibility in this regard. In general, it is suggested that a lowland river with 20 to 25% plant cover provides optimum conditions for land drainage purposes and for the conservation of aquatic insects and fish (Haslam 1978; Hellawell and Bryan 1982). In Denmark, a flood-safe biomass for aquatic plants is estimated to be between 100 and 200 g m[-2] (Kern-Hansen and Holm 1982). Presuming the latter figure to be applicable under Irish riverine conditions it is clear from Fig. 1 (A to F) that many of the rivers examined, and indeed many rivers throughout Ireland, support excessive weed growths throughout most of the growing season.

Having established that weed problems exist, it is important to identify the cause of the problem and control or eliminate this. As the likely causes often include leaching of fertilisers, pollution from point sources, drainage, and water abstraction among others, instant cures are rarely available. One must therefore, in the interim, focus on the most appropriate and

H

environmentally responsible weed control procedures which are available.

Before deciding upon a control strategy it is critical that the nuisance plant species is correctly identified, preferably to the level of species. A failure to correctly identify the target plants, regardless of the control procedure selected, could exacerbate the weed situation. For example, with mechanical or manual control it is important that the cut is synchronised with a particular stage in the life cycle of the target plants if growth is to be suppressed rather than stimulated (Soulsby 1974; Zonderwijk and van Zon 1978). With chemical control a range of plant species will react differently to different herbicidal treatments. Thus, fennel pondweed and water crowfoot, plants which are often difficult to distinguish *in situ*, will react very differently to treatment with Midstream (Fig. 1E and F). Even where environmental control, in the form of bankside shading, is applied, shade-sensitive species will be suppressed while shade-tolerant species will continue to be problematic. It is therefore important that those administering or advising in relation to weed control procedures have more than a casual knowledge of aquatic plants and their growth strategies.

Having correctly identified the problem plant species, the only realistic options for prompt control in Irish rivers are mechanical or chemical treatments. Based on the preliminary results from the present trials it is clear that weed cutting provides only short-term control. In fact, the results of these and other studies have shown that cutting may actually stimulate plant growth, thereby producing a greater maximum standing crop than would otherwise have been produced (Brooker and Edwards 1975; Dawson 1978; Johnstone 1982). Thus, if it is desirable to maintain a relatively weed-free channel over the full growing season two or three cuts may be required. This method of control can therefore be expensive, both in terms of capital outlay for machinery and in labour costs. However, the advantages of mechanical weed control lie in the fact that it is quick, the cut can be localised to certain areas, and species diversity is largely maintained (Caffrey 1989).

Chemical control offers a number of advantages over mechanical control, the most important perhaps relating to the cost of treatment. Harbott and Rey (1981) found the cost of chemically controlling submerged macrophytes in drainage channels to be 33% cheaper than mechanical means. Comparative figures for chemical and mechanical treatment on Irish canals show the former to be 14% cheaper than the latter (Office of Public Works, personal communication). Another advantage relates to the speed and ease with which the chemicals can be applied, and the fact that the immediate environmental impact is often less severe than that caused using mechanical means (Caffrey 1986).

The results presented in Fig. 1 (C, D and F) demonstrate the high level of control which is possible using herbicides against broad-leaved pondweeds and water crowfoot. The results of one spraying in April/May maintained these aquatic plants at manageable levels for the full growing season. Additionally, it was possible to partially treat the river, thereby leaving sufficient vegetation in place to provide for the needs of aquatic fauna and wildfowl. Two plant species which did not respond to chemical treatment, with

Midstream, were broad-leaved pondweed and fennel pondweed (Fig. 1E and F). Further experimentation will be conducted with these species, particularly as the latter is potentially one of the most troublesome weed species in Ireland. Another plant group which will require special analysis in future years is the filamentous green algae. This group of pollution-tolerant species (Caffrey 1986) is extending its range in Irish rivers and, as yet, no suitable controlling agents are available.

In Ireland, those responsible for the maintenance of relatively weed-free channels have been slow to use herbicides for aquatic weed control. This stems from a basic fear that the herbicides used may be toxic to fish or contaminate potable water supplies. However, concern as to risks to the health of consumers or to fish stocks from the proper use of herbicides in water supply catchments are safeguarded by formal clearance procedures (Pieters and de Boer 1971; Tooby 1976; Hellawell and Bryan 1982). The main risks to water supplies come from accidental spillage or human errors in the use of the herbicides.

Based on the results of ongoing experiments with herbicides under Irish conditions it is clear that, when used carefully and with consideration for the environment, herbicides are valuable and safe management tools which deserve a place in the armoury of any water manager.

REFERENCES

AN FORAS FORBARTHA 1986 *Water Quality in Ireland. The Current Position.* Dublin. Water Resources Division, An Foras Forbartha.

ANONYMOUS 1986 *Inland Fisheries. Strategies for Management and Development.* Dublin. Central Fisheries Board.

ARTS, W.B.M. and VAN WIJK, I.A. 1978 New developments in the mechanical control of water weeds. *Proceedings EWRS 5th Symposium on Aquatic Weeds*, 1-6. Wageningen. EWRS.

BARRETT, P.R.F. 1978 Aquatic weed control : necessity and methods. *Fisheries Management* 9, 93-101.

BROOKER, M.P. and EDWARDS, R.W. 1975 Aquatic herbicides and the control of aquatic weeds. *Water Research* 9, 1-55.

CAFFREY, J.M. 1986 Macrophytes as biological indicators of organic pollution in Irish rivers. In D.H.S. Richardson (ed.), *Biological Indicators of Pollution*, 77-87. Dublin. Royal Irish Academy.

CAFFREY, J.M. 1988 *Trials to Evaluate the Environmental Impact of Chemical Weed Control, using Casoron and Midstream, on Selected Irish Waters.* Office of Public Works commissioned report. Dublin. Central Fisheries Board.

CAFFREY, J.M. 1989 *Aquatic Plant Management in Irish Canals.* Office of Public Works commissioned report. Dublin. Central Fisheries Board.

DAWSON, F.H. 1978 Aquatic plant management in semi-natural streams: the role of marginal vegetation. *Journal of Environmental Management* 6, 213-21.

DAWSON, F.H. 1981 The reduction of light as a technique for the control of aquatic plants - an assessment. *Proceedings AAB Symposium on Aquatic Weeds and their Control, Oxford*, 157-64. Wellesbourne. Association of Applied Biologists.

DAWSON, F.H. and KERN-HANSEN, U. 1979 The effect of natural and artificial shade on the macrophytes of lowland streams and the use of shade as a management technique. *Internationale Revue der gesamten Hydrobiologie* 64, 437-55.

DE NIE, H.W. 1987 The decrease in aquatic vegetation in Europe and its consequences for fish populations. *EIFAC/CECPI Occasional Paper* 19, 1-52.

FOX, A.M., MURPHY, K.J. and WESTLAKE, D.F. 1986 Effects of diquat alginate and cutting on the submerged macrophyte community of a *Ranunculus* stream in northern England. *Proceedings EWRS/AAB 7th Symposium on Aquatic Weeds*, 105-11. Wageningen. EWRS.

HARBOTT, B.J. and REY, C.J. 1981 The implications of long-term aquatic herbicide application: problems associated with environmental impact assessment. *Proceedings AAB Symposium on Aquatic Weeds and their Control, Oxford*, 219-31. Wellesbourne. Association of Applied Biologists.

HASLAM, S.M. 1978 *River Plants*. Cambridge University Press.

HELLAWELL, J.M. and BRYAN, K.A. 1982 The use of herbicides for aquatic weed control in water supply catchments. *Journal of Institute Water Engineering and Science* 56, 221-33.

JOHNSTONE, I.M. 1982 Waterweed decline: a resource allocation hypothesis. *Proceedings EWRS 6th Symposium on Aquatic Weeds*, 45-53. Wageningen. EWRS.

JORGA, W., HEYM, W. and WEISE, G. 1982 Shading as a measure to prevent mass development of submersed macrophytes. *Int. Revue ges. Hydrobiol.* 67, 271-81.

KERN-HANSEN, U. and HOLM, T.F. 1982 Aquatic plant management in Danish streams. *Proceedings EWRS 6th Symposium on Aquatic Weeds*, 122-31. Wageningen. EWRS.

MINISTRY OF AGRICULTURE, FISHERIES AND FOOD 1979 *Guidelines for the Use of Herbicides on Weeds in or near Watercourses and Lakes*. Booklet 2078, MAFF Publications, Middlesex.

PIETERS, A.J. and DE BOER, F.G. 1971 Dichlobenil in the aquatic environment. *Proceedings of the EWRS 3rd International Symposium on Aquatic Weeds*, 183-93. Wageningen. EWRS.

SEAGRAVE, C. 1988 *Aquatic Weed Control*. Surrey. Fishing News Books.

SOULSBY, P.G. 1974 The effect of a heavy cut on the subsequent growth of aquatic plants in a Hampshire chalk stream. *Journal of the Institute Fisheries Management* 5, 49-53.

STOTT, B. 1977 On the question of the introduction of grass carp (*Ctenopharyngodon idella* Val.) into the United Kingdom. *Fisheries Management* 8, 63-71.

TOOBY, T.E. 1976 Effects of aquatic herbicides on fisheries. *British Crop Protection Council Monograph* 16, 62-77.

ZON, J.C.J. VAN 1977 Grass carp (*Ctenopharyngodon idella* val.) in Europe. *Aquatic Botany 3*, 143-55.

ZONDERWIJK, P. and VAN ZON, J.C.J. 1978 Aquatic weeds in the Netherlands : a case of management. *Proceedings EWRS 5th Symposium on Aquatic Weeds*, 101-4. Wageningen. EWRS.

Dr Charles I. Essery (University of Ulster at Jordanstown, Newtownabbey, Co. Antrim): Shading was mentioned briefly in the introduction. Do you have any quantitative evidence of the efficiency of this method of control/ management compared with other methods? What is "drainage board's" (O.P.W.) view on such methods of management?

J.M. Caffrey: In rivers < 15 m wide, shading with bankside bushes and trees can effectively reduce the biomass of submerged macrophytes. No quantitative estimation of the efficiency of shading has been produced under Irish conditions, although Dawson and Kern-Hansen (1979) and Jorga, Heym and Weise (1982) have attempted to quantify the effects on stream vegetation. The one undesirable side-effect from shading relates to the accumulation of leaf litter in the river.

The O.P.W. are currently adopting bankside shading as a weed control strategy.

Dr Paul Giller (Dept. Zoology, University College Cork): Herbicides are far more acceptable than dredging which has catastrophic effects on invertebrates, fish and mammals (like otters), but while you were happy with short-term effects of herbicide control, any such control has longer-term indirect effects on invertebrate communities and fish by altering the habitat structure of the system. This may in turn have knock-on effects upstream and downstream of the controlled stretch.

J.M. Caffrey: The herbicides selected for use in aquatic systems are generally those which exert partial or localised weed control. When used discriminately, these herbicides can effectively remove target weed from designated sections of channel while leaving the untreated marginal vegetation intact. This vegetation provides a habitat for invertebrates and fish and helps to maintain macrophytic diversity. In this manner, only the minimum habitat disturbance is caused.

Mike Fitzsimons (Shannon Reg. Fisheries Board, Thomond Weir, Limerick): Many of the weed problems illustrated would appear to be in drained rivers where the natural flow regime has been disrupted, coupled with total removal of bank shading, and the problem of increased nutrients. If proper rehabilitation was completed would this remove the necessity of this type of weed control in the river situation?

J.M. Caffrey: The judicious rehabilitation of drained rivers might not completely resolve aquatic weed problems in these channels, but it would certainly diminish the problem. Drained channels are commonly uniform, open, slow-flowing and silted, and provide instream conditions which favour the proliferation of nuisance aquatic plant species. Judicious rehabilitation will alter the flow regime, create a pool/glide/riffle sequence and introduce bands of trees (deciduous) along the banksides, the shade effect of which will reduce plant growth. The problem of nutrient enrichment cannot be resolved through rehabilitation and must be tackled at source.

Dr Don Cotton (Regional Technical College, Sligo): If you were to successfully control macrophytes for one or more years, do you think that a subsequent relaxation of chemical control is likely to lead to a serious resurgence of the problem owing to the loss or suppression of herbivores from the system and the increased availability of nutrients in sediments?

J.M. Caffrey: Where chemical control is the primary weed management procedure in a watercourse, results have shown that cessation of treatment for even one year can result in a resurgence of the weed problem. This reflects the fact that the chemicals applied are generally non-residual and are undetectable in the water or sediment within 4-5 months post-treatment. As such, there is no residual carry-over from one year to the next. However, the best results are achieved where an integrated weed control policy is operated. This involves alternating control procedures so that areas which are chemically treated in Year 1 will be mechanically treated in Year 2 and vice versa. Where weed control through bankside shading is possible, this offers the most environmentally safe procedure.

In: Steer, M.W. (ed.) 1991 *Irish Rivers : Biology and Management,* pp 99-112. Royal Irish Academy, Dublin.

MANAGEMENT OF MIGRATORY TROUT (*Salmo trutta* L.) POPULATIONS IN THE ERRIFF AND OTHER CATCHMENTS IN WESTERN IRELAND

Martin M. O'Farrell

Department of Zoology, Trinity College Dublin

and

Ken F. Whelan

Central Fisheries Board, Mobhi Boreen, Glasnevin, Dublin

ABSTRACT

The exploitation of migratory trout (*Salmo trutta* L.) by rod and line in the Erriff and Cashla catchments is described. Over four years (1984-7) exploitation rates on Tawnyard Lough averaged 13.1% (C.V.%=38). In the Cashla catchment the average weight of migratory trout taken on rod and line was positively correlated with altitude of capture location.

Results of kelt and immature sea run fish census work carried out over four years (1985-1988) at the outflow from Tawnyard Lough on the Erriff system are presented. The numbers of kelts (including immature sea run fish) recorded each year averaged 9.1 ha⁻¹(of lentic water upstream of the trap site) (C.V.%=16.3).

Migratory trout smolt output from three sub-catchments in 1988 (Tawnyard on the Erriff, School House Lakes on the Cashla and Lough Fee on the Culfin) are compared. The number of smolts recorded averaged 26.8 ha⁻¹ (C.V.%=89.6).

The management of migratory trout recreational fisheries is discussed in terms of angling regulations (imposition of size limits and slot size limits) and stock enhancement methodology (addition and maintenance of spawning gravels, modification of stream nursery areas, regulation of stream discharge, hatchery operation).

INTRODUCTION

Went (1962) reviewed the extensive literature on Irish migratory trout (*Salmo trutta* L.) and work has been carried out on the majority of catchments in Ireland which support migratory trout, e.g. literature cited in Fahy (1978a). However, with the exception of the long-term research of the Salmon Research Trust of Ireland (SRTI) on the Burrishoole system (Fig. 1), all of these investigations have been essentially qualitative. Le Cren (1985) summarised much of the recent British and Irish work on migratory trout. The Central Fisheries Board initiated the River Erriff Research Programme in August 1983 with a view to the development of management strategies for Atlantic salmon (*Salmo salar* L.) and migratory trout stocks and the recreational fisheries based on these stocks. This paper outlines some of the findings of this programme and also incorporates recent research results from other catchments in the west of Ireland.

Latterly there has been considerable concern among fishery proprietors regarding the status of migratory trout stocks in Connemara and Mayo. This concern has been brought about by a significant decline in rod catches in most Connemara fisheries since 1985. In 1988 a "Sea Trout Action Group" composed of various fishing interests, both state and private, was formed with the intention of investigating this apparent decline in migratory trout stocks.

STUDY AREAS

Research on migratory trout was carried out in three catchments in western Ireland (Fig. 1). The location of the Burrishoole River system where the SRTI has been carrying out quantitative research on migratory trout for over 30 years is also shown.

The Erriff catchment has an area of 166.3 km^2 (Fig. 2). It is an acid spate river system. The underlying rock derives from the Ordovician period. The recreational fishery on the system is renowned. On the river (identified with a dashed line in Fig. 2) Atlantic salmon is the main quarry, migratory trout being a by-catch, while on Tawnyard Lough the reverse is the case. The surface area of all lakes in the catchment amounts to 189 ha (1.14% of total catchment area). Tawnyard Lough (surface area 55 ha or 0.3% of total catchment area) is the only lake on the system where migratory trout are taken on rod and line. The downstream trap site on the afferent stream from Tawnyard Lough is labelled X in Fig. 2.

The Cashla catchment has an area of 82.2 km^2 (Fig. 3). It is an acid system with many lakes connected by stream channels of varying character. The underlying rock is granite. The recreational fishery on the system is renowned. Throughout the catchment, migratory trout is the main quarry, Atlantic salmon being a by-catch. The surface area of all lakes in the catchment amounts to 496 ha (6.03% of total catchment area). Migratory trout enter most of these lakes though fishing is concentrated on those identified in Fig. 3. The downstream trap site on the afferent stream from the School House Lakes is labelled X in Fig. 3.

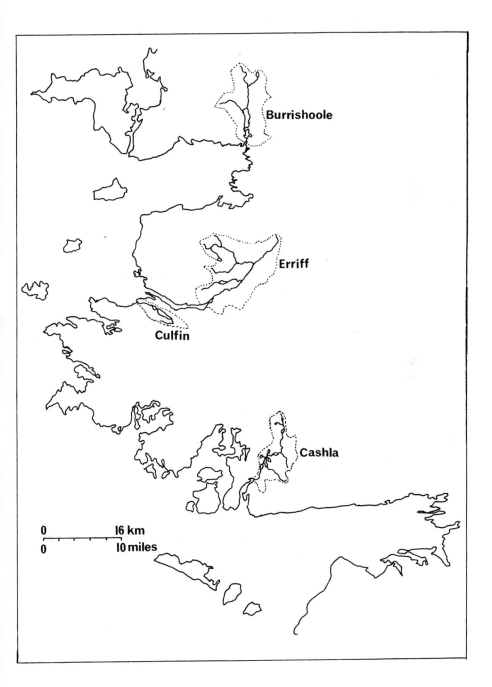

Fig. 1. The locations of the Cashla, Culfin, Erriff and Burrishoole catchments in the west of Ireland.

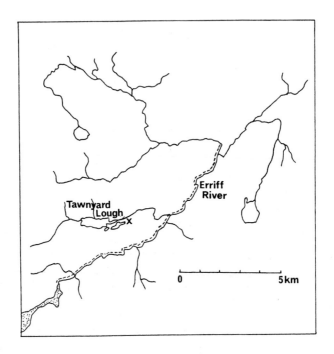

Fig. 2. The Erriff catchment showing Tawnyard Lough and the Erriff River (dashed line indicates the extent of the rod fishery). The downstream trap site is labelled X.

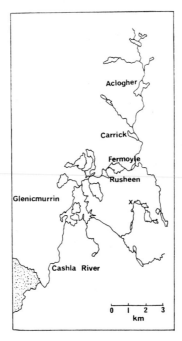

Fig. 3. The Cashla catchment showing the main fishing lakes. The downstream trap site is labelled X.

The Culfin catchment has an area of 22.7 km^2 (Fig. 1). It is an acid system with two large lakes occupying a central position in the catchment. The underlying rock is predominantly metamorphic schist and gneiss. The surface area of all lakes in the catchment amounts to 210.2 ha (9.3% of total catchment area).

MATERIALS AND METHODS

The terminology for migratory trout recommended by Allan and Ritter (1977) was used throughout. Immature sea run fish which overwintered in freshwater and were recorded in the downstream traps are included with true kelts (migratory trout which have recently spawned) in all calculations.

Migratory trout taken on rod and line in the Erriff Fishery (River Erriff and Tawnyard Lough) during the 1983-7 angling seasons and from the Costello and Fermoyle Fishery (Cashla system) during the 1987 angling season were measured for fork length (mm) and weighed to nearest 7 g (1/4 oz divisions). A sample of scales was also taken from each fish from an area between the dorsal and adipose fins and above the lateral line. Using designs similar to those of Wolf (1950) and Cresswell (1977), downstream traps were used in the capture of migratory trout smolts and kelts on afferent streams from Tawnyard Lough (1985-8), School House Lakes (1988) and Lough Fee (1988). The duration of trapping operations at each site was as follows:

Tawnyard Lough: 11 March - 26 April 1985,
16 January - 6 June 1986,
16 December 1986 - 5 June 1987,
28 October 1987 - 1 June 1988;

School House Lakes: 8 April - 23 May 1988;

Lough Fee: 15 April - 25 May 1988.

In the calculation of maximum possible and estimated exploitation rates it was assumed that all migratory trout which entered Tawnyard Lough in any year did so before the end of the angling season and that all surviving kelts were successfully counted downstream. In the estimation of kelt and smolt densities the numbers recorded at each trapping location represent the minimum output.

Captured fish were anaesthetised using a 20% solution of quinaldine (2-methylquinoline) in acetone to give a quinaldine concentration of 15 mgl^{-1}, measured for fork length (mm) and scaled as above. Scales were interpreted using a microfilm reader.

RESULTS

Rod catch analyses

Erriff River. The sea age structure of the rod catch for the 1983-7 angling seasons is presented in Table 1. The 0-sea-winter (or 0 sea age) group predominated in samples. The mean sea age for all years was 0.20 (C.V.% = 35.0).

Table 1. Sea age structure (%) of Erriff River migratory trout rod catch (1983-1987).

Sea Age	1983 (n=350)	1984 (n=421)	1985 (n=506)	1986 (n=244)	1987 (n=310)
0	87.4	87.2	82.8	84.0	76.1
1	10.8	11.4	12.6	12.3	17.7
2	1.2	0.9	3.4	3.3	5.5
3	0.6	0.2	1.0	0.4	0.6
4	0	0	0.2	0	0
Mean sea age	0.15	0.14	0.23	0.20	0.30

Tawnyard Lough. The sea age structure of the rod catch for the 1983-7 angling seasons is presented in Table 2. This structure varied between years. For example, the 0 sea age group accounted for 66.6% of the sample in 1985 compared with 29.5% in 1987. The mean sea age for all years was 0.74 (C.V.% = 32.4).

Table 2. Sea age structure (%) of Tawnyard Lough migratory trout rod catch (1983-1987).

Sea age	1983 (n=28)	1984 (n=57)	1985 (n=93)	1986 (n=110)	1987 (n=44)
0	57.1	54.4	66.6	43.6	29.5
1	25.0	36.8	22.6	41.8	34.0
2	7.1	5.3	5.4	10.9	31.8
3	3.6	3.5	4.3	3.6	4.5
4	7.1	0	1.0	0	0
Mean sea age	0.78	0.58	0.50	0.74	1.11

Cashla system. Statistics were collected on 1,123 migratory trout taken in the Costello and Fermoyle Fishery during the 1987 angling season. In Table 3 the average weight (g) at each capture location is presented. The data show that the mean weights of migratory trout taken on rod and line increase with altitude of capture location.

Table 3. Statistics on migratory trout taken in the Cashla system in 1987.

Location	Altitude (m)	Number	Mean weight (g)	S.E.	C.V.%
Cashla R.	14.0	87	353.8	40.8	107.7
Glenicmurrin L.	28.3	738	346.0	9.5	74.5
Rusheen L.	32.0	52	316.5	27.1	61.8
Fermoyle L.	35.7	168	435.3	23.2	69.2
Carrick L.	74.7	36	526.4	52.3	59.5
Aclogher L.	113.7	42	578.6	68.3	76.5

Kelt census work

Table 4 describes the sea age structure of migratory trout kelts recorded. The duration of trapping varied each year, as did the numbers of migratory trout kelts recorded (mean number = 503; C.V.% = 16.2). Over the four years for which data are available the mean number of migratory kelts recorded per hectare of lentic habitat upstream of the trap site was 9.13 (C.V.% = 16.2).

The sea age structure of the kelts (aged as previous year's rod-caught fish) varied considerably between years. For example, 0 sea age fish which entered the lake in the summer of 1985 comprised 66.9% of the kelts recorded migrating downstream in 1986, while the 0 sea age fish which entered the lake in the summer of 1987 comprised only 43.1% of kelts recorded migrating downstream in 1988. Over the four years for which data are available the mean sea age was 0.63 (C.V.% = 23.8).

Table 4. Sea age* structure (%) of migratory trout kelts (1985-1988).

Sea Age	1985 (n=412)	1986 (n=499)	1987 (n=489)	1988 (n=610)
0	55.5	66.9	55.8	43.1
1	29.8	21.0	33.9	35.7
2	12.1	8.4	7.9	16.7
3	1.7	2.6	2.0	3.8
4	0.7	1.0	0.2	0.5
5	0	0	0	0.2

Mean sea age	0.62	0.50	0.57	0.84

* Aged as previous year's rod-caught fish.

Table 5 shows the actual numbers of kelts derived from individual smolt classes. In Table 5 kelts recorded in 1988 include the 1987 0 sea age group, the 1986 1 sea age group etc. This table also shows that the 1985 smolt class was much stronger than the 1982-4 smolt classes.

Table 5. Numbers of kelts from 1982-7 smolt classes recorded in Tawnyard Lough downstream trap.

			Smolt class			
Sea age	1982	1983	1984	1985	1986	1987
0	-	-	229	334	273	263
1	-	123	105	166	218	-
2	50	42	39	102	-	-
3	13	10	23	-	-	-
4	1	3	-	-	-	-
5	1	-	-	-	-	-

Exploitation rates

An attempt was made to calculate the exploitation rates of migratory trout by rod and line on Tawnyard Lough during the 1984-7 angling seasons (Table 6). A maximum possible exploitation rate was calculated by relating the numbers of migratory trout taken in any year to the numbers of kelts counted downstream the following year. Thus maximum possible exploitation rates averaged 17.6% (C.V.% = 36.9). Estimated exploitation rates were calculated which assumed that 70% (based on findings at the SRTI) of the migratory trout which were present in the lake at the end of the angling season survived to the kelt stage. Estimated exploitation rates averaged 13.1% (C.V.% = 38.2).

Table 6. Migratory trout rod exploitation rates (%) on Tawnyard Lough (1984-1987).

Year	1984	1985	1986	1987	1988
Rod catch	108	137	125	52	42
No. kelts	-	412	499	489	610
Max. % exploit.	20.8	21.5	20.3	7.9	-
% Exploitation*	15.5	16.1	15.2	5.6	-

* Assuming 70% overwinter survival of all kelts.

Smolt output

Table 7 describes the results of the migratory trout smolt census work in terms of lentic area upstream of trap sites and the numbers, density and biomass of smolts counted downstream at each site. The mean number of smolts recorded was 26.8 ha^{-1} (C.V.% = 89.6).

Table 7. Details of migratory trout smolt census work carried out at three locations in 1988.

	Location of trap and grid. ref.		
	Tawnyard L93 68	School Hse M03 31	Fee L78 62
Lentic area (ha)	55.0	33.9	187.6
Altitude (m)	47.8	98.1	46.9
No. smolts	2877	800	866
Mean Wt (g)	75.9	58.9	-
No. smolts ha^{-1}	52.3	23.6	4.6
Kg smolts ha^{-1}	3.9	1.4	-

DISCUSSION

Catchment characteristics

Fahy (1978b) recognised that Connemara catchments where migratory trout were the main quarry of the fisherman had characteristically a significant amount of lentic habitat. Thus the Erriff (with 0.3% of the catchment area consisting of lentic habitat accessible to migratory trout) is primarily a salmon fishery while the Cashla (with 6.0% of the catchment area consisting of lentic habitat accessible to migratory trout) is primarily a migratory trout fishery. Lentic habitat in the Burrishoole River system (Fig. 1) accounts for 5.7% of the catchment area and as comparison will later be made between results obtained in this catchment and those obtained in the present paper this catchment characteristic is noteworthy. The importance of lentic habitat may be confined to the provision of suitable holding areas for fresh run fish, the provision of suitable recruitment areas for juveniles or the provision of suitable wintering areas for spent and immature sea run fish. Armstrong (1974) also recognised the importance of freshwater lentic habitat as a wintering area for both spent and immature anadromous Dolly Varden (*Salvelinus malma*).

Stock structure and strength

Migratory trout have such variable life histories in terms of smolt age, age at first maturity, number of spawnings etc. that it is convenient to think of them in terms of sea age groups. Thus fish which have spent two summers feeding at sea and have completed one winter since migrating to sea are termed 1-sea-winter or one sea age group fish. The mean sea age of rod-caught fish on the Erriff River was less than that of rod-caught fish on Tawnyard Lough. Similarly the average weight of sea trout taken on the Cashla system in 1987 increased with increasing altitude of capture location. In the Burrishoole system the average sea age of migratory trout taken on Lough Furnace (a tidal lough) is less than that of fish taken on Lough Feeagh (C.P.R. Mills, personal communication). The reason for this occurrence is probably the fact that 0 sea age fish (the majority of which are immature) tend to occupy the lower reaches of catchments when they re-enter freshwater and they are also probably easier to catch on rod and line than older fish. The fact that there is good agreement between the mean sea age of fish taken on rod and line on Tawnyard Lough and that of the kelts counted downstream the following year indicates that once migratory trout have homed each sea age group is exploited by anglers in proportion to its relative abundance.

Previous attempts to describe the sea age structure of migratory trout populations from scale samples taken from rod-caught fish must therefore be cautiously interpreted as no attempt was made to identify the location within a catchment from which samples were collected.

Between-year differences in the sea age structure of the population may reflect variations in the strength of smolt classes, in the survival of fish at sea, in the percentage of 0 sea age fish which ascend to Tawnyard Lough as 0 sea age fish or in the overwinter survival of kelts.

There is no evidence from the data examined (1985-8) that the stock of migratory trout in Tawnyard Lough is declining. The 1987 estimated exploitation rate of 5.6% is low when compared with estimates for the previous three years.

Smolt output

The main concern of the fishery manager is the maximisation of smolt output. The fate of migratory trout at sea is outside his control. Thus all his effort is concentrated on managing migratory trout and their progeny in freshwater. Data presented in Table 7 indicate that there is considerable variation in smolt output between sub-catchments from different catchments. However, the smolt output from these sub-catchments has been compared only in terms of lentic habitat upstream of the trap sites and there may be additional characteristics (e.g. spawning and lotic (running water) nursery area, altitude, bathymetry of lentic habitat etc.) of importance. Quantitative information on smolt output is available for Lough Feeagh on the Burrishoole system in Mayo (Fig. 1). For the years 1985-8 the smolt output averaged 8.5 ha^{-1} (C.V.%=12.6) (Annual Report of the Salmon Research Trust of Ireland, No. XXXII) while the figure for 1988 was 9.5 ha^{-1} (C.P.R. Mills, personal communication). Thus the minimum smolt output per hectare from Tawnyard Lough in 1988 was 5.5 times that of Lough Feeagh.

The relationship between the resident and migratory components of the trout population in catchments producing migratory fish is not clearly understood at present. Dellefors and Faremo (1988) found from mark-recapture experiments carried out in two streams in south-western Sweden that pre-migratory mature males exhibited a reduced tendency to migrate to sea compared to pre-migratory immature fish. There is also evidence from the Cashla catchment (O'Farrell, unpublished) that significant numbers of resident females mature without migrating to sea. Mills (personal communication) has also found that resident trout in the Burrishoole system dominate the trout population numerically and also make a significantly greater contribution than migratory trout to total trout egg deposition each year. In the Erriff system no 0 sea age fish containing residual ova have been recorded (O'Farrell, unpublished). All this evidence indicates that trout which migrate to sea are immature fish. There is no precise information available at present on the proportion of trout which mature and remain resident.

Management

The cost of effective management of migratory trout recreational fisheries in the west of Ireland is significant in relation to revenue from rod letting, with the result that some fisheries are unmanaged while others are managed only during the angling season. Management is largely confined to the voluntary implementation of a size limit and a compulsory fly-only rule. The effect of applying various size limits on the Erriff Fishery was investigated in 1985 (Table 8).

Table 8. Potential impact of various size limits on migratory trout rod catches in the Erriff Fishery in 1985.

Size limit	Location	% of catch to be returned
10 inches (25.4 cm)	Erriff River	9.2
11 inches (27.9 cm)	Erriff River	46.3
12 inches (30.5 cm)	Erriff River	75.6
10 inches (25.4 cm)	Tawnyard Lough	2.1
11 inches (27.9 cm)	Tawnyard Lough	22.7
12 inches (30.5 cm)	Tawnyard Lough	57.4

Thus a size limit of 12 inches (30.5 cm) would protect the majority of 0 sea age fish but would also seriously impact on the catch. The vexed question of 0 sea age group conservation has been approached in various manners. Piggins (1975) argued that 0 sea age fish should be retained as their contribution to the rod catch as 0 sea age fish far outweighed what their contribution would be as one sea age or older fish. Evidence attesting to the survival of released 0 sea age fish is also lacking. O'Farrell et al. (1989), after consideration of the sex ratio, percentage maturation, fecundity and relative abundance of each sea age group, calculated that sea age groups made the following contribution to ova deposition in the Tawnyard Lough sub-catchment in 1986 (ova deposition estimated at 252,140 from a spawning escapement of 745 fish).

Sea age	0	1	2	3	4	5
% Contribution to ova deposition	5.6	40.6	35.3	13.0	4.4	1.0

Thus 0 sea age fish made a negligible contribution to ova deposition in the Tawnyard Lough sub-catchment in 1986 and therefore should not be conserved. On the other hand, one and two sea age fish deposited 75.9% of all ova. A slot size limit of 35-45 cm (13.8-17.7 inches) would protect these fish while allowing the capture of 0 sea age and larger trophy fish.

Estimated exploitation rates of migratory trout on Tawnyard Lough averaged 13.1% (C.V.%=38.2) for the years 1984-7. On Lough Feeagh exploitation rates for the years 1970-81 averaged 11% (Mills et al. 1986) while for the years 1985-7 they averaged 8.8% (C.V.%=43.2) (Annual Report No. XXXII, Salmon Research Trust of Ireland). Both these fisheries are regulated and only fly-fishing is allowed, as is the case on most migratory trout fisheries in the west of Ireland. These low exploitation rates are unlikely to impact seriously on stocks. Thus current levels of exploitation and of angler education indicate that size limits or slot size limits are unnecessary or impractical or both. However, should migratory trout stock levels decrease significantly in future these management practices may deserve further consideration.

The management of migratory trout populations should also entail an understanding of the nature of migrations. Armstrong (1974) found that non-spawning anadromous Dolly Varden commonly entered streams other than their natal streams and overwintered in lakes. O'Farrell (unpublished data) has recorded immature 0 sea age fish from the Delphi Fishery, south Mayo, which were marked at the smolt stage as they migrated from Tawnyard Lough. The extent to which 0 sea age fish overwinter in non-natal catchments is presently not understood and requires further investigation.

Enhancement of stock

Enhancement of migratory trout stocks in the west of Ireland has centred on addition to and cleaning of spawning gravels, regulation of discharge from lakes, protection of sea run fish at sea and in freshwater, and the operation of enhancement hatcheries either for the production of unfed fry, feeding fry or smolts. While the success of enhancement with unfed or fed fry has never been evaluated, that of enhancement with reared smolt has. The Burrishoole results of smolt enhancement (Mills et al. 1989) showed that many released smolts did not migrate to sea and that returns to the fishery and contribution to the rod catch did not justify the smolt rearing costs involved.

The dynamics of migratory trout progeny in lotic habitats are now well understood. Elliott (1984) found that the initial density of ova was the dominant factor affecting the subsequent density of 0+ trout in each cohort and that variations in this initial density accounted for a high proportion of the variation in production between cohorts. He found that density-dependent mortality during the establishment of feeding territories was the principal population regulator and that zero group trout which failed to establish and defend territories were poorly conditioned, i.e. their weight was usually well below that expected for a resident fish of equal length. Elliot also found that the drought summers of 1969, 1976 and 1983 had a significant impact on cohort biomass and production owing to decreased growth rates during these summers caused by high water temperatures and reductions in available drifting food and stream habitat. Are these findings applicable to catchments where lentic habitat predominates?

Chadwick and Green (1985) studied Atlantic salmon smolt output from a largely lacustrine watershed where water covered 17% of the catchment and 98.5% of this was lentic habitat. Smolt census work near the mouth of the system provided an estimate of salmon production for the entire watershed while an electrofishing survey quantified production in lotic areas. The difference was assumed to reflect salmon production in lentic habitats. These workers found that while production m^{-2} of lentic habitat was only 3% of that of lotic habitat, over 67% of total production occurred in lentic habitats.

Lentic habitats may be of paramount importance to juvenile trout. Many observers have noted the occurrence of fry and parr in lake margins and concluded that the lentic habitat represented an important nursery area. Thus the occurrence of a significant amount of lentic habitat in a catchment may reduce the impact of drought conditions on juvenile trout populations.

The operation of a hatchery with on-growing facilities with a view to planting summer parr directly into lentic habitat is worthy of consideration as it bypasses a critical period (Elliott 1989) of density-dependent mortality in lotic habitat and may contribute significantly to trout density in the lentic habitat.

ACKNOWLEDGEMENTS

The authors wish to thank all those who contributed to the River Erriff Research Programme and to the collection of data in other catchments. The support of the proprietors of the Costello and Fermoyle Fishery is gratefully acknowledged.

REFERENCES

ALLAN, I.R.H. and RITTER, J.A. 1977 Salmonid terminology. *Journal du Conseil. Conseil permanent international pour l'exploration de la mer* 37, 293-9.

ARMSTRONG, R.H. 1974 Migration of anadromous dolly varden (*Salvelinus malma*) in southeastern Alaska. *Journal of the Fisheries Research Board of Canada* 31, 435-44.

CHADWICK, E.M.P. and GREEN, J.M. 1985 Atlantic salmon (*Salmo salar* L.) production in a largely lacustrine Newfoundland watershed. *Verhandlungen der Internationalen Vereinigung für theoretische und angewandte Limnologie* 22, 2509-15.

CRESSWELL, R.C. 1977 A simple and inexpensive trap for catching downstream migrants. *Fisheries Management* 8, 43-6.

DELLEFORS, C. and FAREMO, U. 1988 Early sexual maturation in males of wild sea trout, *Salmo trutta* L., inhibits smoltification. *Journal of Fish Biology* 33, 741-9.

ELLIOTT, J.M. 1984 Growth, size, biomass and production of young migratory trout *Salmo trutta* in a Lake District stream, 1966-1983. *Journal of Animal Ecology* 53, 979-94.

ELLIOTT, J.M. 1989 Mechanisms responsible for population regulation in young migratory trout, *Salmo trutta*. 1. The critical time for survival. *Journal of Animal Ecology* 58, 987-1001.

FAHY, E. 1978a Variation in some biological characteristics of British sea trout, *Salmo trutta* L. *Journal of Fish Biology* 13, 123-38.

FAHY, E. 1978b Performance of a group of sea trout rod fisheries, Connemara, Ireland. *Fisheries Management* 9, 22-31.

LE CREN, E.D. (ed.) 1985 The biology of the sea trout. Summary of a symposium held at Plas Menai 24-26 October 1984. Llanelli, Wales. Atlantic Salmon Trust/Welsh Water Authority.

MILLS, C.P.R., MAHON, G.A.T. and PIGGINS, D.J. 1986 Influence of stock levels, fishing effort and environmental factors on anglers' catches of Atlantic salmon, *Salmo salar* L., and sea trout, *Salmo trutta* L. *Aquaculture and Fisheries Management* 17, 289-97.

MILLS, C.P.R., QUIGLEY, D.T. and CROSS, T.F. 1989 Rearing and ranching of sea trout, *Salmo trutta* L., in the Burrishoole River system. In M.J. Picken and W.M. Shearer (eds), Proceedings of 'The Sea Trout in Scotland', Oban, 18-19 June 1987.

O'FARRELL, M.M., WHELAN, K.F. and WHELAN B.J. 1989 A preliminary appraisal of the fecundity of migratory trout *Salmo trutta* in the Erriff catchment, Western Ireland. *Polskie Archiwum Hydrobiologii* 36 (2), 273-81.

PIGGINS, D.J. 1975 Stock production, survival rates and life-history of sea trout of the Burrishoole River system. *Annual Report of the Salmon Research Trust of Ireland Incorporated*, Appendix 1, No. XIX, 45-57.

WENT, A.E.J. 1962 Irish sea trout, a review of investigations to date. *Scientific Proceedings of the Royal Dublin Society* 1A (10), 265-96.

WOLF, P. 1950 A trap for the capture of fish and other organisms moving downstream. *Transactions of the American Fisheries Society* 80, 41-5.

Part 4

Impact of development on river biology

In: Steer, M.W. (ed.) 1991 *Irish Rivers : Biology and Management*, pp 115-134. Royal Irish Academy, Dublin.

A PROGRESS REPORT ON THE REHABILITATION OF THE KILMASTULLA RIVER, CO. TIPPERARY (1986 - 1988)

John Bracken, Mary Norton and Helen Phillips

Department of Zoology, University College Dublin

ABSTRACT

Opencast mining operations commenced in the mountains over-looking Silvermines village, Co. Tipperary, in the late sixties. Lead, zinc and small quantities of silver were extracted from the crude ore by a process using many toxic substances including cyanide, copper sulphate and a variety of xanthates. During the earlier years the upper half of the system was shown to be polluted by heavy metals which accumulated in the bottom sediments. By 1975 the Zoology Department, University College Dublin, commenced the first of two surveys on the system. This survey lasted for two years and proved that the major problem associated with the system was not caused by heavy metals, as would be expected, but by a bottom deposition of xanthates which obliterated spawning territories, microhabitats for the macroinvertebrate fauna and caused the salmonid populations to move downstream.

Mining operations ended in 1982, at which stage the river was severely polluted (N. Roycroft, pers. comm.). A second, smaller sur-vey was again undertaken by the Zoology Department, U.C.D., in 1986. This paper represents a progress report on the second survey. Where possible, comparisons have been drawn between the two sets of data under the headings of water chemistry, benthic macroinverte-brate community structure, and macrophyte and algal distribution. Fish tissues, bottom sediments and water samples have been anal-ysed for heavy metal contamination.

INTRODUCTION

The discovery in 1967 of substantial deposits of lead and zinc in the mountains overlooking Silvermines village, Co. Tipperary, led to the commencement of opencast mining operations two years later. A processing factory, employing over 500 men, was constructed near the source of the Kilmastulla River, a minor tributary of the Shannon. The Shannon system is the largest inland waterway in Ireland (Fig. 1) and it is controlled and managed by the Electricity Supply Board, Ireland.

Prior to the commencement of the mining operations water samples were collected and analysed regularly by the ESB laboratory staff. Field officers also carried out a quantitative electrofishing programme to determine the standing crop of salmonids. Results showed that the system was a typical Irish salmonid water. The upper reaches, composed of five small tributaries, were ideal for spawning and as nursery areas for young fish. Mr N. Roycroft (pers. comm.) commented that over these years fish, mainly salmonids, were observed moving up and down the system at regular intervals. Salmon, trout and some coarse fish entered the Kilmastulla River during the autumn and winter months to spawn. During the spring and early summer a reverse downstream movement of fish occurred from the tributary into the main Shannon at Parteen.

Over the next six years the Kilmastulla was subjected to a high level of pollution as a variety of effluents entered the river from the main factory site. In 1975 the Zoology Department, University College Dublin, was commissioned to assess the current status of the river. Funding was provided jointly by the Department of Fisheries and the National Board for Science and Technology. Over the next two years detailed information was collected monthly at 18 stations on the system. Physical, chemical and biological monitoring was undertaken (Norton 1978). Each station was analysed for water chemistry, benthic macroinvertebrate community structure and macrophyte distribution. The distribution of algae, the concentration of suspended solids and the rate of deposition of sediment were also investigated. Fish tissues, together with suspended and bottom sediments, selected macroinvertebrates and different macrophyte species, were analysed for heavy metal contamination.

By 1982 mining activities had ceased but the river was still heavily polluted and practically devoid of any fish life (N. Roycroft, pers. comm.). A second limnological survey was initiated in 1986 (Phillips and Bracken) and is still continuing. Although this survey was severely curtailed, owing to financial constraints, the results obtained have been extremely interesting. The present paper attempts to compare and contrast both surveys. The extent of the earlier damage to the macroinvertebrates is compared to the present community structure of the aquatic organisms together with an update on the existing status of the river.

The study area

The Kilmastulla River is one of the smallest tributaries of the Shannon (Fig. 2). Rising in the Silvermine and Arra mountains, Co. Tipperary, it flows for 18.4 km to enter the Shannon at Parteen. The head-waters are formed by a confluence of five small feeder streams. The Erinagh River contributes three of these streams. This is by far its largest tributary.

Fig. 1. Location map showing Shannon system.

The Kilmastulla then flows, mainly in a westerly direction, through rich agricultural farmland (Fig. 1). Riffle areas make up approximately 10% of the total but there are many glides and deep pools which are ideal as holding areas for salmonids. It has great potential as a salmonid water. A total of 12 stations have been surveyed (Table 1).

The dominant aquatic plants present in the Kilmastulla are *Ranunculus penicillatus* (water crowfoot) and *Potamogeton natans* (broad-leaved pond-weed). Any particular stretch of the river can have up to 80% weed cover. The bank vegetation consists of many mature trees: *Fraxinus excelsior* (ash), *Acer pseudoplatanus* (sycamore) and *Aesculus hippocastanum* (horse chest-nut). The understorey vegetation is dominated by *Anthriscus sylvestris* (cow-parsley), *Urtica dioica* (nettle) and *Hedera helix* (ivy).

Ten species of fish have been recorded for the system including *Salmo salar* L., the salmon, *Salmo trutta* L., the brown trout, *Anguilla anguilla* (L.), the freshwater eel, *Phoxinus phoxinus* (L.), the minnow, and *Gasterosteus aculeatus* L., the three-spined stickleback. These were the commonest spec-ies. Adult salmon were rarely encountered but there were reasonable num-bers of juvenile salmon and trout located throughout the system. Small shoals of minnow and stickleback were located from Crannagh Bridge down-stream (see Fig. 1). Isolated specimens of *Esox lucius* L., the pike, *Perca flu-viatilis* L., the perch, *Neomacheilus barbatulus* (L.), the stone loach, *Gobio gobio* (L.), the gudgeon, and *Lampetra fluviatilis* (L.), the river lamprey, were located mainly in the lower reaches.

Sources of pollution during the mining operations

The ore extraction process involved the use of sodium cyanide, copper sulphate, lime, sodium isopropyl xanthates and potassium amyl xanthates. Tailing ponds were constructed to hold the overspill from the factory. Large volumes of water were extracted from the river and used by this flotation method. Lead, zinc and small quantities of silver were extracted.

Within two years of the mining operations commencing more than half the Kilmastulla system was polluted by heavy metals in the bottom sediment of the river. This damage was minimal, however, when compared to the eutrophic problems subsequently created by a bottom deposition of xan-thates. Xanthates covered the entire bed of the river to a depth of 5-6 cm, from the Yellow River to a point immediately below Station 7. The macroin-vertebrate communities were decimated while most fish moved downstream.

The situation was exacerbated by the addition of untreated sewage from the general factory area also entering the system. Mining activities ceased in 1982.

MATERIALS AND METHODS

Physical and chemical measurements

During the 1975-77 survey monthly water samples were collected at 18 sites along the Kilmastulla River (Fig. 1). This was drastically curtailed

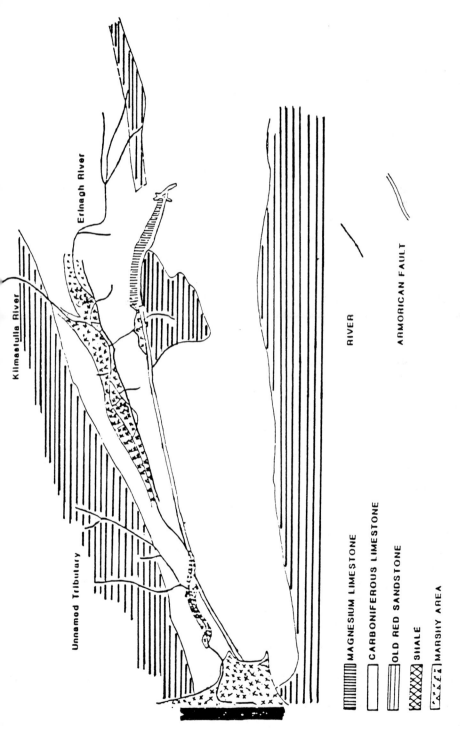

Fig. 2. Summary of geology, Kilmastulla River system catchment area.

Table 1. Kilmastulla River: chemical analyses (1972-7 and 1983-8).

STATION	Period (year)	D.O. O₂ % sat.	B.O.D. 5 Day Test (O₂) mg/l	Ammonia NH₃-N mg/l	Nitrate NO₃-N mg/l	Nitrite NO₂-N mg/l	Oxidised* Nitrogen mg/l
1 Erinagh Bridge (R 814 725)	1972-77	93.2	2.18	0.087	0.13	<0.003	-
	1983-88	88-100	0.5-1.2	0.016-0.03	-	-	0.28-1.1
2 Above Effluent A (R 804 726)	1972-77	76-109	1-2.8	0.01-0.179	0.14-2	0.0003-0.16	-
	1983-85	95	1.2	0.03	-	-	0.94
3 Below Effluent A (R 804 726)	1972-77	68-109	1.4-11	0.02-1.89	0.06-1.7	0.002-0.13	-
	1983-85	98	1.2	0.04	-	-	1.0
4 Above Effluent B (R 809 707)	1972-77	87-110	0.6-2	0.01-0.18	0.02-0.36	0.0003-0.17	-
	1983-87	90	0.6-1.0	0.02-<0.1	-	-	0.1-<0.7
5 Below Effluent B (R 806 718)	1972-77	81	0.6-2.26	0.22-2.8	0.05-4.9	0.004-0.33	-
	1983-87	90	0.8-1.0	0.03-0.1	-	-	0.1-0.9
6 Crannagh (R 786 718)	1972-77	44-108	1.2-16	0.03-1.85	0.07-3.5	0.006-0.4	-
	1983-88	100-105	0.26-1.4	0.033-0.2	-	-	0.89-2.3
7 Phil Kelly's house (R 755 710)	1972-77	77.5	3.5	0.576	0.06	0.004	-
	1983-88	106	0.5-1	0.025-0.1	0.28-1.8	0.002-<0.1	-
8 Dublin Road Bridge (R 726 697)	1972-77	78	1.96	0.7	0.11	0.004	-
	1983-88	106	0.03-1	0.016-0.1	0.181-1.6	0.001-<0.1	-

Table 1 (continued).

STATION	Period (year)	D.O. O₂ % sat.	B.O.D. 5 Day Test (O₂) mg/l	Ammonia NH₃-N mg/l	Nitrate NO₃-N mg/l	Nitrite NO₂-N mg/l	Oxidised* Nitrogen mg/l
9 Killaloe Road Bridge (R 710 693)	1972-77 1982-88	81-116 100-104	1.6-3.2 0.43-1.3	0.01-2.6 0.025-<0.1	0.1-2 -	0.005-0.2 -	- 0.84-<1.5
10 Hatchery (R 680 675)	1972-77 1987-88	86 105	1.29 1-2	0.49 0.04-<0.1	0.17 0.16-1.4	0.006 0.001-<0.1	- -
11 Effluent A outfall (R 803 726)	1972-77 -	32-92 -	2.9-37	0.16-5.7 -	0.14-3.8 -	0.003-0.34 -	- -
12 Effluent B outfall (R 808 718)	1972 1975-77	70-120 75	1.6-6.5 2.8	0.43-3.0 0.685	0.4-4.3 0.12	0.003-0.07 0.005	- -

* Nitrate and Nitrite are given together as Oxidised Nitrogen for 1983-1985. These measurements were done by *An Foras Forbartha*

N.B. : All results are mean values or a range of mean values over a period specified.

during 1988 to 6 sites sampled monthly for the April to September period. Fourteen standard physico-chemical parameters were measured during the studies (Table 2). Although the methods used differed over the course of time any standard error due to this difference may be considered negligible.

Benthic macroinvertebrate sampling

Quantitative macroinvertebrate sampling was carried out during both surveys by using a Surber sampler (area $1/16$ m^2). Duplicate samples were taken at each station and sampling was confined to riffle areas. Sweep samples for adult insects were also taken during the summer months on the surrounding vegetation to augment the faunal data. Faunal samples were subsequently sieved and sorted in the laboratory and stored in 5% formalin. Oligochaetes were cleared in Amman's lactophenol. A solution of 10% KOH was used to clear the chironomid larvae which were then mounted in Canada Balsam.

Fish sampling

Each time fish were captured in the Kilmastulla River the same electro-fishing techniques were employed. The apparatus consisted of an electrical A/C generator (220/230 volts) and a metal frame dipnet (mesh size 7.5 mm) attached to the unit by an electric cable, thus allowing at least 30 m of river to be sampled at a time. As fish were captured the salmonids were retained and transferred to plastic containers. The numbers of other species were recorded and the fish released. Three stations were electrofished and standing crop estimates of the salmonid populations were calculated.

Each trout and salmon parr captured was examined individually, having been first anaesthetised by MS222 for ease of handling, for length in cm and weight in g (a Pescola scales was used to weigh each fish). A small number of scales were removed from the shoulder region of each fish and retained for age analysis. All fish above 80 mm were then stomach-flushed. Foster's (1977) modification of the Seaburg (1957) technique was used. It involved the use of a pressurised flow of water (9.09 l/min.) from a gusher galley pump connected to polythene tubing. Tubing of 2.5 mm inside diameter and 3.0 mm outside diameter was used for trout less than 10.0 cm; for fish between 10 and 15 cm tubing of 3.5 mm inside diameter and 4.0 mm outside diameter was used, while for fish larger than 15 cm the tubing had an inside diameter of 4.0 mm and a 5.0 mm outside diameter. A collecting funnel directed the stomach contents into a storage vial and at the same time allowed excess water to escape. The operation could be carried out on anaesthetised fish in less than 0.5 min. Removal efficiencies ranged from 98% to 100% (Kelly-Quinn and Bracken 1988).

Each fish was carefully revived in well-oxygenated water before being returned to the river. A 100% survival rate was thus achieved.

Table 2. Kilmastulla River: water metal analyses (1972-7 and 1983-8).

STATION	Metals							
	Cu(ppb)		Cd(ppb)		Zn(ppb)		Pb(ppb)	
	1972-1977	1987	1972-1977	1987	1972-1977	1987	1972-1977	1987
1 Erinagh Bridge	36.7	NS	<0.0	NS	23.7	NS	2.5	NS
2 Above Effluent A	11	NS	<0.1	NS	42.3	NS	8.2	NS
3 Below Effluent A	33.3	NS	1.1	NS	69.6	NS	16.2	NS
4 Above Effluent B	27.1	8	5.9	1.6	644.1	250	35.7	297
5 Below Effluent B	11.1	7	1.3	5.7	171.4	1190	13.7	73
6 Crannagh	28.7	<2	1	0.4	91.6	160	9.6	5.4
7 Phil Kelly's house	19.3	<2	0.6	<0.4	37.7	140	4.3	4.2
8 Dublin Road Bridge	16.3	<2	<0.1	<0.4	70.8	100	5.6	2.8
9 Killaloe Road Bridge	14.7	<2	0.6	<0.4	60	90	4.3	3
10 Hatchery	12.9	<2	0.5	<0.4	65.8	120	2.3	2.4
11 Effluent A outfall	91	NS	10.7	NS	315	NS	90	NS
12 Effluent B outfall	21.7	NS	4	NS	263.2	NS	32.1	NS

N.S. = Not significant

Heavy metal analysis

The four heavy metals studied were lead, zinc, copper and cadmium. Small numbers of salmonids, together with water samples, macroinvertebrates and sediments, were retained for detailed analysis.

In the case of the fish, six fish tissues including the liver, lateral muscle, kidney, eye, gill and gonad were examined. The procedure involved drying the tissues at 110 °C for 48 hours, then adding conc. HNO_3 - 10 ml to 0.5 g weighed tissue, after which heat was applied until all the tissue was solubilised. Once solubilised the heat was increased to 80° - 110 °C until the solution became straw-coloured. Then 1 ml of peroxide was added and the solution allowed to digest at 100 °C for 30 min. After this the solution was allowed to evaporate until only a white precipitate remained. Finally, this was made up to 10 ml in 4% HNO_3. The solution was then read on the atomic adsorption spectrophotometer. Water samples and sediment were also examined for heavy metals.

RESULTS

Physical and chemical analysis

By the time the 1975-1977 survey commenced, effluents from the tailing ponds had already had severe toxic effects on the river. Dissolved heavy metals and organic chemicals were responsible for this damage. There was a major reduction in dissolved oxygen in the river water. Sediments with a high xanthate content covered the substratum and obliterated most of the available microhabitats. The detrimental effect of the heavy metal deposition was clouded, therefore, by the presence of these organic chemicals. The symptoms produced were similar to those caused by cultural eutrophication. The high B.O.D. values and the critical reductions in oxygen during September 1976 and May 1977 in the system reflected this problem. Recent studies clearly demonstrate that this problem has been rectified naturally by the system ridding itself of these contaminants since the mining activities ceased. Presently all values for B.O.D. are < 1.0 and the mean D.O. concentration is 104%. The only station still suffering from the ill-effects of the accumulation of sediment on the river bed is Crannagh Bridge. This section, however, has been canalised and the flow regimes altered. Chemical results are included in Table 1.

Effluent A (Station 11) increased the concentrations of ammonia, nitrite and sulphate entering the river. The mean ($\bar{x}$) ammonia concentration at this station was 2.396 mg/l N (1975-1977) and a maximum concentration of 5.68 mg/l N was recorded during the 1972 survey. The later studies showed a dramatic decrease in ammonia concentrations. All studies, however, have recorded concentrations of ammonia in excess of those recommended for salmonid waters. In the light of other less stringent limits, however, it is clear that ammonia is no longer a threat to the system. Klein (1966) maintains that concentrations of > 0.2 mg/l NH_3 provide presumptive evidence of polluted water and An Foras Forbartha (Flanagan and Toner 1972) stipulate that levels of 1 mg/l NH_3 are the threshold values in relation to fish kills.

When the mine was in operation nitrite levels also rose in the river below the effluent discharges and, as with ammonia, showed no appreciable decrease with distance from source. Concentrations of nitrite in the region of 0.01 mg/l NO_2 are considered the threshold values for salmonid waters. There is reason to believe that silage and particularly slurry may be entering the river from adjacent farms. Thus, this parameter will have to be carefully monitored in the future.

Considerable amounts of sulphate were added to the system by the effluents, effluent A in particular. At no stage, however, have sulphate concentrations in the Kilmastulla been a cause for concern.

The values attained by dissolved metals during the 1972 survey were on occasion considerably above acceptable limits. The more extended survey 1975-1977 revealed, however, that dissolved metals in the river were generally not significant from the pollutional aspect.

The river already had high background levels of heavy metals. This was confirmed by the fact that the macroinvertebrates at Station 1, upstream of the mine effluents, frequently possessed equal or greater metal concentrations than those living in the lower polluted parts of the river. Analysis of the bottom sediments confirmed that mining operations substantially raised the total metal concentrations above the natural background levels. High concentrations of particulate zinc and lead, and to a lesser extent copper and cadmium, entered the river by the effluents. The heavy metals underwent dilution, however, and 6 km below the effluent outfall were undetectable.

Clearly the situation in relation to heavy metals in solution has improved. The concentrations of all four metals, copper, cadmium, zinc and lead, have decreased. The 1972 and 1975-1977 studies revealed values in the range ND - 240 p.p.b. and 1 to 118.4 p.p.b. respectively for copper. The 1987 investigation showed all readings to be < 2 p.p.b.

Similar decreases in cadmium concentrations also brought this metal within permissable loading levels (see Table 2). All stations on the Kilmastulla now register < 0.4 p.p.b. Cd, whereas previously readings of 20.2 p.p.b. were recorded.

Concentrations of zinc are currently well below the mandatory limit of 330 p.p.b. set out for salmonid waters. The concentrations attained by lead were never a cause for concern. All values recorded for the Kilmastulla River in February 1987 were below 6 p.p.b. Pb. Even during the 1975-1977 investigation values never exceeded 57.4 p.p.b. Pb.

Water samples were taken from April - September 1988 for heavy metal analysis. These were analysed by flame atomic absorption spectrophotometry (A.A.S.). Neither copper, cadmium nor lead were detected. These samples will be dried down, reconstituted in a smaller volume and subsequently analysed by either flame or graphite furnace A.A.S. as required. This may possibly reveal a further decline in heavy metals in solution.

126

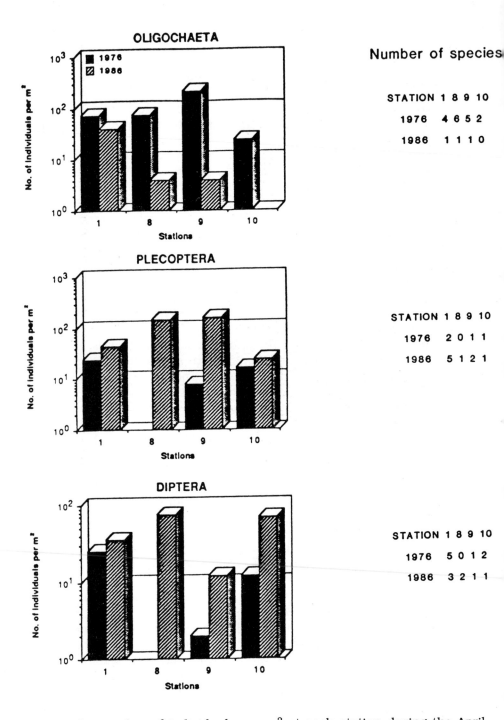

Fig. 3. The number of individuals per m² at each station during the April 1976 and May 1986 investigations. Total numbers are further broken down into the number of species present at each station for both studies.

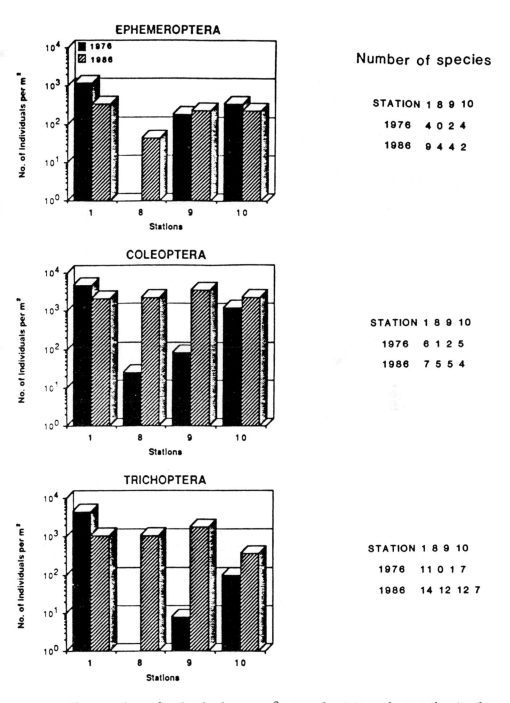

Fig. 4. The number of individuals per m² at each station during the April 1976 and May 1986 investigations. Total numbers are further broken down into the number of species present at each station for both studies.

Benthic macroinvertebrate studies

When the mine effluent was entering the Kilmastulla, the increase in bottom sediments was very important in determining the distribution of the benthic macroinvertebrates, as was the reduction in dissolved oxygen. The extent of the change in community structure between 1975-1977 and 1986-1988 is dealt with here, with particular reference to the quantitative sampling of April 1976 and May 1986.

Six invertebrate groups have been chosen to show the improvement in conditions in the Kilmastulla (see Figs 3 and 4). These are the Oligochaeta, Ephemeroptera, Plecoptera, Coleoptera, Trichoptera and Diptera. An analysis of variance was performed on total abundance, species diversity and each of the major faunal groups. This revealed no statistically significant differences at the 0.05 level. From this, it is concluded that uniform conditions now prevail along the course of the river. This is in marked contrast to the earlier years when the mine effluent was discharging. Then, the river was divided into six categories including clean, polluted, and recovery areas. A species list is included in Table 3.

The trend since April 1987 is one of improved species diversity and increased faunal abundance. The faunal abundance and species diversity of the Ephemeroptera, Trichoptera and Plecoptera have improved in the ten years since the last survey. During the previous investigation, the clean water association was recorded only above the point discharges. Moderately intolerant species of the clean water assemblage returned to the fauna near the mouth of the river.

The Ephemeroptera, which are known to be pollution-sensitive, were abundant at all the stations sampled. There was no statistically significant difference between the mean number of individuals occurring at any of the different stations. Ten species in all were recorded. As mentioned previously the Ephemeroptera form a component of the clean water assemblage. This assemblage is comprised of the Ephemeroptera, Plecoptera, Trichoptera and Elminthidae. All respire by gills and live in waters with high dissolved oxygen concentrations. *Baetis rhodani* is known to be less affected than other species of Ephemeroptera. When the Kilmastulla suffered from organic pollution, this tolerant species was the most abundant. It occurred throughout the system but was more numerous at the clean stations. *Rhithrogena semicolorata* (Curt.), on the other hand, is confined to running waters and is known to have a high oxygen demand. It was recorded during the present study; its occurrence during the earlier study was extremely rare. The presence of two species of *Ecdyonurus* and the absence of *Ephemera* and *Caenis* (1 specimen recorded) confirm the nature of the river bed. The Ecdyonuridae require clean stones for attachment. Conversely *Ephemera* and *Caenis* species are known burrowers in oxygenated muds. The species of Ephemeroptera found thus reflect the stony nature of the substratum.

Looking at the Trichopteran data for the periods 1975-1977 and 1986-1988, changes in distribution are readily seen. The genus *Agapetus* comprised 79% of the total Trichopteran fauna recorded during the 1987 study. This genus formed 15.8% of the population at Station 1 during the earlier study, but it was virtually non-existent below the effluent discharge. The Polycentropodidae are capable of withstanding mild organic pollution and *Polycentropus flavomaculatus* (Pict.) occurred in the recovery zone during the earlier study.

The numbers and diversity of Plecoptera were extremely low during the earlier study. Only the most tolerant of the Plecopteran species, *Leuctra fusca* (L.), had a widespread distribution. The 1987 study showed an increase in Plecopteran diversity, but only now in the present study (1988) is the typical clean water abundance and diversity of Plecoptera evident. Nine species were abundant during the present investigation whereas *Leuctra fusca* (L.) was the only species with a widespread distribution during the earlier study.

The Coleoptera were the most abundant of all the groups recorded in the Kilmastulla. The Elminthidae are the most important component of the Coleopteran fauna, indicating clean water conditions. One specimen of *Brychius elevatius* was recorded from Station 1; this species is also particularly sensitive to pollution. *Dytiscus marginalis* L. is capable of withstanding the oxygen-depleting effect of organic pollution, owing to adaptations of the respiratory system. Two species of *Dytiscus marginalis* were recorded, one at Station 1 and the other at Station 4. Neither *Brychius elevatius* or *Dytiscus marginalis* can be regarded as indicator organisms; their abundance is too low. The abundance, diversity and distribution of the Elminthidae are certainly an indication of clean water conditions.

The Diptera were very poorly represented in the Kilmastulla River. Comparing this study with that of April 1976 we see the most startling discrepancy. The large proportion of Diptera found in the earlier samples corresponds with a high abundance of Chironomidae. The Chironomidae, and particularly the Chironomini, are very tolerant of polluted conditions. They have large quantities of haemoglobin in their blood and thus can withstand levels of deoxygenation. During the present study the Chironomidae and Ceratopogonidae were not abundant. Both groups were studied from a qualitative rather than quantitative viewpoint. Seven genera were found, five of which belonged to the subfamily Orthocladiinae. The Orthocladiinae require well-oxygenated water and are generally found in less polluted systems. The other Chironomidae found belonged to the subfamilies Tanypodinae and Chironomini. It must be stressed, however, that no family of the Chironomidae was well represented, indicating the clean nature of the water and the lack of a suitable substrate. Other Diptera species recorded were five in number. Of these *Psychoda* and *Ephydra* spp are regarded as most tolerant organisms (Flanagan and Toner 1972). Their numbers were, however, too low for their presence to have any significance as indicator organisms.

Table 3. Kilmastulla River: macroinvertebrate species list.

ORDER	FAMILY	GENUS/SPECIES
Eulamellibranchiata	Pisididae	*Pisidium* sp.
Oligochaeta	Enchytraeidae	spp indet.
Hydracarina	Sperchonidae	*Sperchon* spp
	Lebertiidae	*Lebertia* sp.
	Torrenticolidae	*Torrenticola* sp.
Amphipoda	Gammaridae	*Gammarus pulex* (L.)
Ephemeroptera	Baetidae	*Baetis rhodani* (Pict.) *Baetis muticus (L.)* *Baetis* sp.
	Heptagenidae	*Heptagenia* sp. *Rhithrogena semicolorata* (Curt.)
	Ephemerellidae	*Ephemerella ignita* (Poda) *Ephemerella notata* (Etn.)
Plecoptera	Taeniopterygidae	*Brachytera putata* (Newm.)
	Nemouridae	*Amphinemura standfussi* (Ris.) *Amphinemura sulcicollis* (Steph.) *Protonemura meyeri* (Pict.) *Protonemura praecox* (Mort.)
	Leuctridae	*Leuctra hippopus* (Kemp.) *Leuctra inermis* (Kemp.) *Leuctra nigra* (Ol.)
Coleoptera	Elminthidae	*Elmis aenea* (Mull.) *Esolus parallelepipedus* (Mull.) *Limnius volckmari* (Panz.) *Oulimnius tuberculatus* (Mull.) *Riolus cupreus* (Mull.)
Trichoptera	Rhyacophilidae	*Rhyacophila dorsalis* (Curt.) *Rhyacophila obliterata* (McL.)
	Glossosomatidae	*Glossosoma boltoni* (Curt.) *Glossosoma conformis* (Neboiss.)

Table 3. (continued).

ORDER	FAMILY	GENUS/SPECIES
	Hydroptilidae Hydropsychidae	*Agapetus delicatulus* (McL.) *Agapetus fuscipes* (Curt.) spp indet. *Diplecrona felix* (McL.) *Hydropsyche fulcipes* (Curt.) *Hydropsyche pellucida* (Curt.) *Hydropsyche siltalai* (Dohl.) *Hydropsyche instabilis* (Curt.)
	Psychomyiidae	*Psychomyia pusilla* (Fbr.)
	Limnephilidae	*Limnephilus centralis* (Curt.)
	Molannidae	spp indet.
Diptera	Simuliidae	*Simulium* sp.
	Chironomidae	spp indet.
	Ceratopogonidae	spp indet

The Oligochaeta were the only class whose representation was favoured by the incidence of pollution. They were the most abundant group at the grossly polluted stations. In the recovery zones the Chironomidae replaced the Oligochaeta as the dominant group. The low abundance and poor distribution of the Oligochaeta may now be ascribed to the improvement in the dissolved oxygen and the nature of the substratum. The large proportion of Diptera found in the earlier samples corresponds with a high abundance of Chironomidae. The comparative data for the April 1976 and May 1986 studies (Fig. 3) do not include the Chironomidae but reveal a greater abundance of other Dipteran species now. No species, however, is sufficiently abundant to have any significance as an indicator organism.

Fish studies

Ten species of fish have been recorded from the Kilmastulla over both surveys. There has been a dramatic increase, however, in the numbers of salmonids present in the system since 1982. Their distribution has also improved greatly. Both of these observations are directly correlated with habitat improvement, improved feeding facilities resulting from an upsurge in the macroinvertebrate communities and improved water quality.

Table 4. Kilmastulla River: salmonid stock densities (no./m²)

| Sampling date | STATIONS | | | | | |
	Erinagh Bridge St. 1	Crannagh Bridge St. 6	Phil Kelly's house St. 7	Dublin Rd. Bridge St. 8	Killaloe Bridge St. 9
April 1988	-	0.029	0.014	-	-
May 1988	0.084	-	0.043	0.069	-
June 1988	0.220	-	0.031	0.060	-
July 1988	0.136	-	0.027	0.029	-
August 1988	0.076	-	0.073	0.048	-
September 1988	0.092	-	0.107	0.045	0.110

Standing crop estimates, recorded by the ESB since the early sixties, showed conclusively that there was a good number of brown trout and young salmon present in the system prior to the commencement of the mining operations. Less than a year after the ore processing got under way the scene had changed drastically. Approximately half the entire length of the channel was covered with a dark sludge which destroyed the microhabitats of the fauna and caused extensive mortality amongst the aquatic organisms. The fish, however, were able to move downstream as water quality deteriorated.

In September 1976 the salmonid populations of the River Nenagh were compared with the survival figures in the Kilmastulla. A stretch 100 m in length was electrofished in each river. The Nenagh was selected on the basis that it flowed over the same geological strata, and was approximately the same width and depth. Results showed that the salmonid population in the unpolluted Nenagh River was ten times higher than that in the Kilmastulla.

The recent fish data for the Kilmastulla show a vast improvement. The results are summarised in Table 4.

DISCUSSION

The results of the physico-chemical and the benthic macroinvertebrate analyses show a dramatic improvement in conditions in the Kilmastulla River. Previously the faunal communities were dominated by species normally associated with eutrophic conditions. These have now been replaced by a clean water assemblage of benthic macroinvertebrates. In relation to chemical parameters, the previously elevated concentrations of many toxins (ammonia, nitrate, sulphate, copper, cadmium and zinc) have returned to pre-pollution levels. All these changes indicate that the Kilmastulla River system is recovering.

Fish studies during the earlier investigation (1975-1977) revealed good growth rates and high condition factors. The food intake reflected the fauna of the area in which the trout were taken. No major differences were detected in the diet of trout taken from different areas. The unusual features of the population were low stock density, the absence of 0+ fish and the paucity of 3+ fish in the samples.

The elimination of spawning facilities and feeding territories would appear to have been the density-limiting factors. This was the combined effect of extensive siltation of the river bed, man-made organic input in the form of xanthates and increased concentrations of dissolved heavy metals.

Electrofishing operations carried out from June 1987 to September 1988 revealed a vast improvement in stock density. Table 4 shows, however, that the population is still well below the optimum of one fish per m^2. Age, growth, survival and fecundity rates will be estimated for the trout together with feeding studies and heavy metal analysis. The performance of wild versus stocked trout will be carefully monitored. The calculation of production figures will allow a management decision to be reached on the economic and commercial viability of restocking the Kilmastulla River.

ACKNOWLEDGEMENTS

The authors wish to thank the Department of Fisheries and the National Board for Science and Technology for conjointly funding the initial survey from 1975 to 1977. Our renewed thanks to the Department of Fisheries for providing a Fisheries Research Studentship to Miss Helen Phillips to enable us to carry out the second survey which is still under way.

The invaluable assistance received from the Electricity Supply Board, and in particular the advice, help and field assistance received from Mr Noel Roycroft, M.Sc, Mr Paddy Barry, Senior Laboratory Technician, and the late Mr C. O'Flynn, Pollution Officer, is once again gratefully acknowledged. Grateful thanks are also extended to the Hatchery staff for their help with many practical field problems.

REFERENCES

FLANAGAN, P.J. and TONER, P.S. 1972 *The national survey of Irish rivers. A report on water quality.* Dublin. An Foras Forbartha.

FOSTER, J.R. 1977 Pulsed gastric lavage, an efficient method of removing the stomach contents of live fish. *Prog. Fish. Cult.* 39, 166-9.

KELLY-QUINN, M. and BRACKEN, J.J. 1988 Brown trout, *Salmo trutta* L., production in an Irish coastal stream. *Aquaculture and Fisheries Management* 19, 65-9.

KLEIN, L. 1966 *Control : River Pollution,* Vol. 3. London. Butterworths.

NORTON, M. 1978 Final report on the Kilmastulla River 1975-1977. Confidential report prepared in the Zoology Department, UCD, on behalf of the Department of Fisheries.

SEABURG, K.G. 1957 A stomach sampler for live fish. *Prog. Fish. Cult.* 19, 137-9.

In: Steer, M.W. (ed.) 1991 *Irish Rivers : Biology and Management*, pp 135-150.
Royal Irish Academy, Dublin.

INVERTEBRATES AS POLLUTION INDICATORS IN IRISH RIVERS

John Lucey

Environmental Research Unit, Regional Water Laboratory, Kilkenny

ABSTRACT

The larger invertebrates of rivers and streams by their nature are good indicators of pollution. These communities show characteristic changes in density and composition with different degrees of organic enrichment. Schemes of water quality classification, based on the known tolerances of the various groups and species, have been developed in which complex biological data may be reduced to a single numerical value for reporting which is comprehensible to the non-biologist. A scheme suitable for Irish rivers has, since its inception 20 years ago, continued to be used, largely unchanged, to assess the biological quality of rivers and streams nationally. A case study of a pollution incident is illustrated to show how important historical data from monitoring programmes can be in assessing the effects of a pollutant. Biological monitoring, using invertebrates, to assess water quality is set to continue for the future in Ireland.

INTRODUCTION

Nature of invertebrates

The invertebrates of Irish rivers and streams comprise many different types ranging from small single-celled organisms to the largest animals, the crayfish and mussels (Lucey and McGarrigle 1987). When the term invertebrate is applied without qualification, however, it generally refers to those visible to the naked eye, and the microscopic forms, such as the protozoans, are usually treated separately. These larger invertebrates, which are known as macroinvertebrates, include insects (nymphs and larvae of some, as well as beetles and bugs), worms (e.g. leeches and flatworms), molluscs (snails, limpets and bivalves) and crustaceans (e.g. water slater and shrimp) among others, and have been defined as those that are retained by a net or sieve with an aperture of 0.6 mm (Weber 1973). In the present context invertebrates will have this meaning.

One estimate of the number of invertebrate species occurring in freshwater in Ireland has been given at 1,071, of which 790 were insects (McCarthy 1986). The benthic invertebrates, i.e. those which live on, in or near the substratum, of running water include representatives of almost all the taxonomic groups that occur in freshwater (Hynes 1970).

Many river invertebrates have annual life cycles with those producing more than one generation in a year relatively scarce (Hynes 1970). Several species have life-spans of more than a year, including some leeches, and the largest mayfly and the two largest stoneflies take two and three years respectively to grow to full size. Many of the insects only spend part of their life in the river as they require an aerial stage to reproduce and complete the cycle. The crayfish lives for several years while the species with the greatest longevity, the pearl mussel, only reaches sexual maturity at about 12 years (Young and Williams 1983) and may live for over a century.

The invertebrates of rivers and streams can be conveniently separated, based on habitat preference, into two kinds, viz. those that normally inhabit the faster-flowing shallow areas called 'riffles' and those that occur in the slower-flowing deeper areas called 'pools'. It is not fully understood why, but riffle areas in natural streams tend to be spaced at more or less regular distances of five to seven stream-widths apart (Hynes 1970). The riffle-dwellers are, by their structural or behavioural adaptations, able to withstand the scouring action of the current. Unlike fish they are relatively sedentary and thus representative of the areas in which they are found.

Nature of pollution

The types of waste discharged to Irish waters are mainly organic in nature (Toner *et al.* 1986) and hence biodegradable. A river can assimilate a certain amount of organic matter depending on the dilution it affords and its reaeration characteristics, which in turn will depend on whether it is sluggish or fast-flowing. It is when this assimilation capacity is exceeded that problems begin and what is called 'pollution' may become evident.

Organic wastes result from a great variety of sources, e.g. creameries, meat factories, beet-sugar processing, farming (animal slurries and silage) and domestic sewage. Although these effluents vary, their basic property is that they contain unstable compounds which are readily oxidised and so use up the dissolved oxygen in the water (Hynes 1960). The deoxygenation is brought about by the heterotrophic activity of microorganisms, chiefly bacteria, which increase vastly in numbers in response to the rich 'food' supply. Such wastes can also settle on the river bed causing siltation, and others may contain, or lead to the formation of, ammonia which is readily soluble in water. These effects of oxygen depletion, siltation and ammonia toxicity acting individually or in combination can, depending on flow, temperature, and in the case of ammonia the pH of the receiving water, exert changes on the ecology of the river.

INVERTEBRATES AND POLLUTION

The main ecological effects of organic effluents are that they change the biological community composition and abundance of organisms below the discharge point by lowering the dissolved oxygen content, causing siltation, and adding nutritive material which may favour the dramatic increase in numbers of certain organisms (Hellawell 1986). The responses of river macroinvertebrates to organic pollution are well understood and are the best documented of all the taxonomic groups in freshwater (e.g. Hynes 1960; Hawkes 1962; Warren 1971). The invertebrates associated with the naturally silted areas (depositing substrata) of rivers are more tolerant than those typical of the riffle areas (eroding substrata) as the latter are particularly sensitive to the effects of low dissolved oxygen and silting.

When a large amount of organic matter enters a river a sequence of characteristic changes in the invertebrate fauna occurs with progress downstream of the discharge. If this is strong enough to result in total deoxygenation then no normal river animals will survive and these will be replaced by the 'rat-tailed maggot', *Eristalis* (*Tubifera*), the larva of the hoverfly, which because of its extensible tail-end air tube is indifferent to the anoxic conditions and can feed on the organic mud so long as the depth is no more than about 10 cm.

More usually, however, in grossly polluted conditions where oxygen levels are low, tubificid worms are characteristic and are often the only macrofauna present, with the silted conditions providing an ideal medium for feeding and burrowing and the absence of predators allowing large increases in population to occur (Mason 1981). Protozoa abound and the zone is often colonised by slime growths ('sewage-fungus') where suitable substrates, such as stones and weeds, are present for attachment. Such growths can extend for considerable distances downstream, particularly in times of low water temperature, and have been observed in one Irish river for some 26 km below a discharge (Lucey 1987a).

The midge larva *Chironomus riparius*, sometimes called the 'bloodworm' (sic) because of its coloration, like the tubificidae has haemoglobin[*] in its blood which acts as a carrier when the oxygen tension of the water is low, but unlike these worms cannot withstand severe deoxygenation although it tolerates fairly high concentrations of ammonia (Hynes 1960). This chironomid lives in a tube which is kept oxygenated by the waving motion of its body and this activity apparently aids recovery of the river bed by oxidising the sediments (Westlake and Edwards 1957).

Other chironomids appear as conditions improve further downstream,

----

[*]The haemoglobin (Hb) of Chironomus, unlike that of mammals, occurs free in solution and its affinity for oxygen is about fifty times that of human Hb (Brennan 1981).

and the next zone, with the water slater *Asellus aquaticus* in great abundance, signifies that recovery is under way. Because of the breakdown of the organic matter and the subsequent release of nutrients, particularly phosphate, this zone, which will be less turbid, may contain the filamentous alga *Cladophora*, with which this isopod is commonly associated. These algal growths may cover large areas of the substratum, much like the slime growths can in the grossly polluted zone. The flatworm *Dendrocoelum lacteum*, a predator of *A. aquaticus* (Reynoldson 1978), may also figure in the fauna if conditions are not too severe, as will leeches (e.g. *Erpobdella, Glossiphonia* and *Helobdella*) and tolerant molluscs such as *Physa fontinalis, Lymnaea pereger* and *Sphaerium* spp.

The shrimp *Gammarus duebeni*, the commonest of the gammarid species occurring in Ireland (Reid 1939; Strange and Glass 1979), which can tolerate, but rarely thrives in, heavily polluted conditions, may appear in great numbers as the process of self-purification of the river continues. The *Gammarus* zone will exhibit a much increased variety of invertebrates with caddis fly larvae (Trichoptera) and nymphs of some species of mayfly (Ephemeroptera) also beginning to occur, with the most tolerant of the latter, viz. *Baetis rhondani* and *Caenis* spp, being the first to appear. Numbers of the net-spinning types, such as the black-fly larva *Simulium* and the caddis *Hydropsyche*, will often increase greatly with slight pollution and the same response is often seen with regard to the latter in rivers which receive plankton-rich water from lake outflows.

Further amelioration and full recovery to clean water conditions will allow the most sensitive mayflies (e.g. *Ecdyonurus* and *Rhithrogena*), including *Ephemera danica*, the largest species which burrows in coarse silt, to join the now highly diverse communities. This last species is synonymous with the word mayfly to the angler and the only one to be recognised as such. Stoneflies (Plecoptera) by their appearance will also signify high water quality and again there is some gradation with regard to sensitivity within the group, with *Amphinemura sulcicollis* followed by *Leuctra* spp showing most tolerance. The largest and most sensitive of the stoneflies, the carnivorous *Perla bipunctata*, may be represented in the fauna where the substratum is suitable but will not always be found in the lower reaches of rivers. The other perlid *Dinocras cephalotes* is also very sensitive to the effects of organic pollution but is much less commonly distributed in Irish rivers (Costello 1988) and hence will not feature regularly among the fauna.

The sequence of characteristic invertebrate fauna changes associated with decreasing levels of pollution as outlined above should be viewed as a model illustrating the sensitivities or otherwise of the groups and species and the succession may not always be so simple and clear-cut in practice. The two interrelated trends which can be recognised when increasing organic enrichment occurs are (Hellawell 1986):

> (i) a reduction in the diverse macroinvertebrate community characteristic of clean water, in which many species are represented by relatively few individuals, towards a condition in which under the influence of severe pollution a few species are represented by very large

numbers of individuals, viz. those species which are able to take advantage of the changes which the pollutant induces and to exploit the increased food supply provided;

(ii) the progressive decrease in abundance of particular indicator species until very few remain and their replacement by species not previously present, or at least not abundantly present.

The changes induced by organic pollution can be said to generally favour the pool-dwelling invertebrate types to the detriment of the riffle fauna. These former types will, under the changed circumstances, be able to extend their range to include the riffle areas.

Observations by biologists in the field and from laboratory experiments have led to the compilation of tolerance lists of groups and individual species upon which systems of water quality classification have been based. The rationale for using macroinvertebrates in preference to other components of the river biota in such schemes has been outlined previously (Lucey 1987b).

APPLICATIONS OF INVERTEBRATES AS INDICATORS OF POLLUTION

In river studies the application of invertebrates as indicators of water quality falls mainly into two areas. Firstly, the most common use is in regional or national surveys of water quality, i.e. monitoring, where many sites on a large number of rivers and streams have to be assessed. Here the data are usually compressed or summarised into a numerical index or other such scheme of classification so that the results can be more easily interpreted by non-biologists. This is important also if the results are to be reported along with other numerical values, such as physico-chemical data, for overall water quality management. Secondly, in conservation or assessment studies of the effects of specific discharges or physical disturbance the raw data, in the form of species lists, may be used directly. The two applications may, however, become linked, such as when a pollution incident occurs on a river and then historical data when available from monitoring will add greatly to the assessment of effects on the biota.

Biological monitoring

A system of biological water quality classification suitable for Irish rivers, similar to contemporary schemes in other countries, was devised by Toner in 1970 and later described (Flanagan and Toner 1972a). The scheme, called the Quality Rating System, has been used since 1971 when An Foras Forbartha carried out the first national survey of river and stream quality (Flanagan and Toner 1972b). It relates the diversity and relative abundance of key groups of macroinvertebrates to 5 basic water quality classes: Q5 good quality, Q4 fair quality, Q3 doubtful quality, Q2 poor quality and Q1 bad quality. To give the scheme continuity the intermediate ratings 1-2, 2-3, 3-4 and 4-5 are also used where appropriate.

In order to simplify the presentation of a large amount of information such as for national overviews (e.g. Water Pollution Advisory Council 1983; Toner *et al.* 1986) the description of water quality is related to three classes:

Class A : unpolluted (Q5, Q4-5, Q4),
Class B : slightly (Q3-4) or moderately (Q3, Q2-3) polluted,
Class C : seriously polluted (Q2, Q1-2, Q1).

A summary of the Quality Rating System, relating the diversity and relative abundance of the indicator groups to the Q values, is given in Table 1. To illustrate this a simplified representation of increasing water quality in a river with distance downstream of a waste discharge, as characterised by the indicator groups, is depicted in Fig. 1.

While the Quality Rating System is based mainly on macroinvertebrate communities, other factors are also taken into account in making the assessment, e.g. presence and abundance of plants (macrophytes and algae) and slime growths as well as substratum condition. The full assessment of the site, from sampling to ascribing a Q value, can be completed in the field within about 20 minutes. The sampling method employed will depend on the water depth at the location and the various methods have been reviewed elsewhere (McGarrigle and Lucey 1983). By far the most satisfactory technique is 'kick sampling' where the operator, using a foot, disturbs the area upstream of a hand-held pond net. This is generally carried out for 2-5 minutes and usually supplemented by a similar period spent picking animals from stones to ensure that those with efficient hold-fast mechanisms are represented in the sample. The sample is transferred to a tray containing river water for examination, and a full inventory of the invertebrates and their relative abundance is taken. The relative abundance levels given in Table 1 are rather discretionary and in practice the following are often used: Present 1-4,

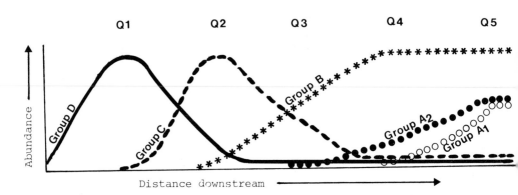

Fig. 1. Schematic representation of water quality changes in a river with distance downstream of a discharge of organic waste as characterised by the invertebrate fauna (see Table 1 for indicator groups and text - Invertebrates and Pollution - for conceptual basis).

Table 1.   Summary of Quality Rating (Q) System relating the diversity (D) and relative abundance of indicator groups to the Q value.

INDICATOR GROUP AND RELATIVE ABUNDANCE

(Q)	A_1 Plecoptera (excluding Leuctra, Ecdyonuridae Ephemera danica	A_2 Ephemeroptera (excluding Baetis rhodani, Cloeon, Caenis, Ephemerella, Trichoptera (cased))	B Leuctra, Baetis rhodani, Cloeon, Caenis, Ephemerella, Gammarus, Trichoptera (uncased), Elminthidae	C Asellus, Chironomidae excluding Chironomus, Hirudines, Mollusca (excluding Physa)	D Chironomus, Physa, Eristalis, Tubificidae and other Oligochaeta	(D)
5	Common	Common	Abundant	Sparse or absent	Sparse or absent	High
4	Sparse or absent	Present	Abundant	Present	Sparse or absent	Slightly reduced
3	Absent	Sparse or absent	Common	Common	Sparse or absent	Significantly reduced
2	Absent	Absent	Sparse or absent	Abundant	Present	Low
1	Absent	Absent	Absent	Sparse or absent	Abundant	Very low

L

Frequent 5-20, Common 21-50, Numerous 51-100 and Dominant 100+. The whole scheme is applied with some latitude and the hydrobiologist employing it may use previous knowledge of a site or an upstream control when evaluating the invertebrate response and ascribing a Q value. Hynes (1959) was of the opinion that attempting to adhere to a rigid system in the application of biological methods tended to lead to them falling into disrepute.

The riffle areas are sampled in preference to other biotypes as the communities therein are most sensitive to the effects of organic pollution. Such areas in most Irish rivers, with perhaps the exception of a large length of the Shannon and the lower reaches of the Barrow, are fordable in the lower summer flows and hence amenable to pond net sampling.

The scheme was designed primarily to detect organic pollution but when toxic influences are suspected the suffix '0' is added to the biotic index (e.g. 2/0, 1/0). The overall ecological consequences of toxic discharges are the reduction in numbers of species (diversity) present and usually in the total number of individuals (abundance) which may lead to the subsequent increase in population of the more tolerant species owing to less competition or predation (Hawkes 1979). The tolerance of different macroinvertebrate groups to metal pollution may differ from that to organic pollution. In general the molluscs and some crustaceans (Malacostraca such as the crayfish, shrimp and water slater) seem to be most sensitive (Whitton and Say 1975), with the oligochaete worms intermediate (Hellawell 1986) and many insect orders such as Diptera and Trichoptera most resistant (Warnick and Bell 1969). Stoneflies are apparently tolerant of zinc (Hynes 1959).

In a study of the Avoca River, insects were found to dominate the invertebrate fauna at sites with a decrease in pH and an increase in heavy metal content (Reynolds 1986). Judicious observations on the biota by a biologist can point to the effects of other toxic substances. For instance, because pesticides are formulated to control particular target groups one might expect them to be more toxic to certain elements, such as related groups, of the biota if they reach rivers, e.g. insecticides might be expected to affect the insect component while macrophytes might be eliminated or damaged by herbicides.

Studies carried out in the 1970s on levels of metals (Flanagan 1974) and pesticides (O'Donnell 1980) in Irish rivers and streams showed that concentrations were generally low, although higher than background concentrations of metals were found at some locations (Kilmastulla and White Rivers) associated with mining. These measurements were limited in scope and there is a need to establish a comprehensive baseline for toxicants in Irish waters (Toner *et al.* 1986).

Specific studies

When surveys are carried out for specific purposes, less extensive than monitoring programmes, a list of the species found and their relative or absolute abundance can be used, with good effect, to report the results. Examples of such surveys would include investigations of the effects or source of a pollution incident. Such lists are also useful in gauging if a watercourse is sufficiently recovered, in terms of fish food, after a fish kill for restocking. Other potential applications might include those of a non-pollution-related type such as conservation studies and assessments of physical disturbance as in the case of arterial drainage operations. As this paper relates to invertebrates as pollution indicators an example of a case study to illustrate such an application will be given.

A pollution incident which occurred in the upper reaches of the River Barrow on 2 August 1987 resulted in an extensive mortality of fish. Reports of a 'lifeless river' following the event prompted An Foras Forbartha to carry out a biological survey of the stretch of river, on 11 August, to establish the true effects on the fauna. With incidents of this nature where the only visible signs of pollution are dead fish it is useful to begin the examination where these have been found in the knowledge that it will be downstream of the causative agency. If the invertebrate fauna have been affected at that point then it is usual practice to proceed in an upstream direction, sampling at intervals, until the effects have ceased. In this way it is sometimes possible to isolate where the pollutant had entered the river. Many dead fish had been collected from the river following the incident, by angling clubs and other interested parties, but many others were still *in situ* in various stages of decomposition. Sampling was begun at Kilnahown Bridge (see Fig. 2) where dead trout (*Salmo trutta*) and sticklebacks (*Gasterosteus aculeatus*) as well as live young minnows (*Phoxinus phoxinus*) were observed. It was possible in this instance to consult previous fauna lists for the sites to be examined in the stretch, as the river had been sampled many times in the past as part of the national surveys of rivers, with the most recent being undertaken on 26 July 1986. The fauna lists for the 1986 survey and this examination would be directly comparable (see Table 2) as seasonal differences, in the life cycle of some insect members of the communities, would be negligible owing to the closeness of the times of year of sampling.

At the first location examined the most sensitive elements of the invertebrate fauna were absent, and upstream at Garryhinch Bridge, where dead trout and sticklebacks were again noted, only a single specimen of each of the mayflies *Ecdyonurus* and *Rhithrogena* were present. Further upstream at Barranagh's Bridge the fauna composition showed only very minor change between the two years with the sensitive elements well represented. This response would indicate *prima facie* that the location was above the area of pollution effects but sampling should always be continued to at least another upstream location to ensure that this is indeed the case. The presence of dead stoneloach (*Noemacheilus barbatulus*) at the site as well as the knowledge that the confluence of the Owenass River was just upstream led to suspicions, and the examination at the next location, Bay Bridge, confirmed these, with disruption to the fauna again apparent and with dead trout, eels (*Anguilla anguilla*), sticklebacks and stoneloach as further ratification. It was concluded that the little change at the previous site was attributable to the large volume of relatively clean water from the Owenass either acting as a buffer to the effects of the pollution there or supplying the sensitive species to the area when the pollution effects had abated. Faunal changes since 1986 were also observed at the next two locations sampled, Two Mile Bridge and Ballyclare Bridge, where sensitive types were missing. Dead fish at these locations included trout, eels, sticklebacks and lamprey (*Lampetra fluviatilis*). At the next upstream site, Ford South of Rearyvalley, the macroinvertebrate response was similar to that obtaining in the 1986 survey with live fish also noted. To ensure that the upstream limit of pollution effects had been reached, sampling was carried out upstream at Tinnahinch Bridge where the invertebrate fauna indicated good water quality.

The conclusions drawn from the survey were that the disruption to the biota had occurred in the area between Ford South of Rearyvalley and Ballyclare Bridge, with the effects evident at least as far as Kilnahown Bridge.

A tributary stream which enters the Barrow between those two upper locations was subsequently examined in the same way at the request of the local authority with responsibility for pollution control in the area. That survey indicated a polluted condition for some distance in the stream which was traced to a 'ditch'. This 'ditch' had apparently polluted the stream in its upper reaches for some time; that conditions had become more severe subsequently was borne out by observations of large numbers of the pollution-tolerant *Chironomus* as well as leeches dead in the upper reaches of the stream. Some specimens of this fly larva and the air-breathing *Eristalis* were the only fauna found immediately downstream of the 'ditch'. The conclusion drawn from the survey of the tributary stream was that pollution of a chronic nature had been occurring which was not apparent upstream of the 'ditch', while in the lowermost reach immediately upstream of the Barrow confluence a limited fauna but without a build-up of tolerant types was evident. At the latter site the absence of sensitive or relatively sensitive faunal elements tended to further indicate that a recent pollution event had occurred.

Investigations by the local authority traced the 'ditch' to a premises and legal proceedings were issued under the Local Government (Water Pollution) Act, 1977. After initial litigation the defendants agreed to pay, in an out-of-court settlement, substantial costs and compensation to the local authority.

In reviewing the effects on the fauna in the Barrow it is curious to note that these appeared most pronounced at the lowermost location examined, Kilnahown Bridge, which although it contained stones was the only site that was non-riffle. This site had supported a variety of insect nymphs in the 1986 survey and it could be postulated that under such conditions these were already at the limit of their tolerance.

Water samples taken at one location on the evening of the incident and again the following morning showed initial very high values for biochemical oxygen demand (a measure of deoxygenating power) and ammonia decreasing with time which would be consistent with an episode of organic pollution.

Some crayfish *Austropotamobius pallipes* were killed in the affected areas but many survived and some of these seemed moribund or 'disoriented'. It is unusual to find this cryptic and nocturnal animal so apparent and it could be that they were in some way narcotised by the pollution effects. However, it is also quite plausible that as predators and scavengers they were exploiting the ready food supply in the form of dead fish. Work carried out on the toxicity of ammonia to North American invertebrates and fish found the former to be generally more tolerant.

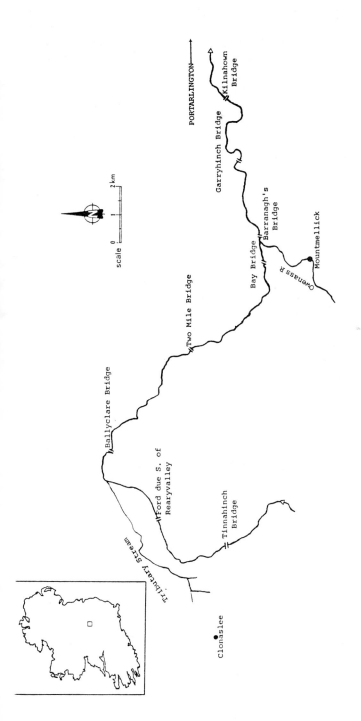

Fig. 2. Outline map of the upper reaches of the River Barrow showing the locations examined on 11 August 1987.

Table 2. Distribution, relative abundance and diversity of macroinvertebrate fauna in the 27 km stretch of the River Barrow between Tinnahinch Bridge and Kilnahown Bridge on 21 July 1986 and 11 August 1987.

Relative abundance levels: Present (P), 1-4, Frequent (F), 5-20, Common (C), 21-50, Numerous (N), 51-100, Dominant (D), 100+

MACROINVERTEBRATE GROUP	Tinnahinch Bridge		Ford S. of Rearyvalley		Ballyclare Bridge		Two Mile Bridge		Bay Bridge		Barranagh's Bridge		Garryhinch Bridge		Kilnahown Bridge	
	'86	'87	'86	'87	'86	'87	'86	'87	'86	'87	'86	'87	'86	'87	'86	'87
PLECOPTERA (Stoneflies)																
Perla bipunctata	F	P	P	F												
Leuctra sp.	F	P	F	F	F			P		P	P	P	F		F	
EPHEMEROPTERA (Mayflies)																
Ecdyonurus sp.	F	C	F	C	F		F		F		P	F	F	P[1]	F	
Heptagenia sp.															P	
Ephemerella sp.	F	F	F	C	C		F		C	P[1]	C	P	C	P[1]	C	
Baetis sp.	F	N	N	D	D	C	D	P	D		D	N	C	F	C	
TRICHOPTERA (Caddis flies)																
Agapetus sp.	P		P		P		F	P	P	P	P	F			P	
Limnephilidae			P	P		P		P	P	P	P		P			
Sericostomatidae	P		P				P		P		P					
Rhyacophilidae			P		P	P	P		P			P				
Hydropsyphidae	F			P	P	P	F		F	F	P	F	P	F		P
Polycentropidae	P		P		P					P	P		P			P
MALACOSTRACA (Crayfish, shrimps & water slaters)																
Austropotamobius pallipes	P	P					P	P	P		P	P	F	P	F	P
Gammarus duebeni						C			F	F	P	C	P	C	P	C
Asellus aquaticus	P		P	P				P	P	P				F	P	P

Taxon																
COLEOPTERA (Beetles)																
Dytiscidae					F	P	P	P	P					P		P
Haliplidae	P		P	P	F	F	F	F	F	F	C	F	F	P	P	P
Elminthidae	P	P	F	F	P	P	P	P	P	P	P	P	P	P	P	P
MOLLUSCA (Snails, limpets and bivalves)																
Potamopyrgus jenkinsi								P	P	P	P	P	P	P		P
Lymnaea peregra			P													
Ancylus fluviatilis			P	P												
Sphaerium sp.					P		P	P	P	P	P	P	P	P		
DIPTERA (Two-winged flies)																
Chironomidae	C	C	N	N	C	C	C	D	C	N	C	C	F	F	C	C
Simuliidae	P	P	P	D	D	N	N	N	N	N	N	C	C	C	F	P
Tipulidae	F	P	F	P	P	P	F	P	P	F	P	P	P	P	P	F
HYDRACARINA (Water-mites)	F		F	C	P	N	F	F	P	C	P	F	P	P		P
HIRUDINEA (Leeches)																
Erpobdella sp.			P	P	P				P	P			P			P
OLIGOCHAETA (Worms)																
Lumbricidae			P													
Tubificidae	P	F	F		F	P	F		P	P	P	P	P	P	P	P
TRICLADIDIA (Flatworms)																
Dendrocoelum lacteum													P			
Diversity	13	15	19	18	14	12	15	13	22	17	19	18	17	16	16	14

[1] Single specimen only.

In that study crayfish were the most resistant and the fingernail clam, the equivalent to *Sphaerium* in the Irish fauna, the most sensitive among the invertebrates (Arthur *et al.* 1987). *Sphaerium* shells were found at four of the affected sites but it was not established at the time if these were live specimens.

In once-off cases of pollution as occurred in this instance, where the discharge is not continuous, a build-up of tolerant species may not result and recruitment of the displaced invertebrates will begin when conditions return to the *status quo ante.* Fish will also return fairly quickly as was noted at the lowermost location examined in the Barrow and in the tributary stream at the confluence where young minnows were recorded. Overall the fish appear to have been more adversely affected than the invertebrates. The survey showed that the effects evident in the river nine days after the incident were far less than the speculative reports of a complete loss of invertebrate fauna as well as fish. It would appear that the most sensitive members of the macroinvertebrate fauna, the stoneflies and mayflies, were the groups affected most (see Table 2), which substantiates the view of hydrobiologists that their tolerance of the effects of organic pollution is similar to that of fish. This study illustrates how important it can be to have previous fauna lists for sites if estimates of the effects of a pollution incident are to be accurately gauged.

FUTURE OUTLOOK

It will be appreciated how the two applications of invertebrates as indicators of pollution, as outlined above, can be interlinked with the information from national surveys providing the necessary background from which the effects of pollution incidents can be established.

The main objectives of the biological monitoring programme are to provide data on the scale of water pollution problems, to gauge the success of remedial measures and to indicate long-term trends which are emerging. Additionally it provides information on species distribution and in the general area of nature conservation.

The programme of national surveys has in the past covered some 2,500 locations on approximately 7,500 km of stream and river length in the state, representing just over 62% of the 12,000 km of channel length delineated on the Ordnance Survey map 'Rivers and their Catchment Basins'. The coverage of the biological surveys has been extended, beginning with the 1987 survey, to include all the rivers and streams shown on the River Catchment map, mentioned above, with the 12,000 km of channel length becoming the long-term baseline for trend detection. The locations are examined on a four-year cycle but with those exhibiting seriously polluted conditions visited annually. The extended national survey covering some 4,500 locations will be fully in place following the 1990 survey.

The practice of issuing annual interim reports on biological monitoring begun by An Foras Forbartha will be continued by its successor, the Environmental Research Unit, and the first of these prepared by the new unit reported on the results of the 1987 surveys (Clabby *et al.* 1988). It is also intended to continue to produce the overviews of the situation regarding water quality nationally as reported in the past (i.e. Clabby 1981; Water Pollution Advisory Council 1983; Toner *et al.* 1986). The assessment of river quality in Ireland has for almost two decades been based largely on biological monitoring, using invertebrates as indicators, and this seems set to continue for the future. The importance of biological surveillance and the use of biological indicators of environmental quality may also increase as a consequence of any future measures implemented in the EC arena. Moves have already been set in train on one front in this regard which might in time lead to the harmonising of methods within the EC region (Lucey 1987b).

REFERENCES

ARTHUR, J.W., WEST, C.W., ALLEN, K.N. and HEDTKE, S.F. 1987 Seasonal toxicity of ammonia to five fish and nine invertebrate species. *Bulletin of Environmental Contamination and Toxicology* 38, 324-31.

BRENNAN, A. 1981 Famous animals and plants. Chironomus. *Biologist* 28, 133-8.

CLABBY, K.J. 1981 *The national survey of Irish rivers. A review of biological monitoring 1971-1979.* WR/R12. Dublin. An Foras Forbartha.

CLABBY, K.J., LUCEY, J. and McGARRIGLE, M.L. 1988 *The national survey of Irish rivers. Biological monitoring. Interim report for 1987.* Dublin. Environmental Research Unit.

COSTELLO, M.J. 1988 A review of the distribution of stoneflies (Insecta, Plecoptera) in Ireland. *Proceedings of the Royal Irish Academy* 88B, 1-22.

FLANAGAN, P.J. 1974 *The National survey of Irish rivers. A second report on water quality.* Dublin. An Foras Forbartha.

FLANAGAN, P.J. and TONER, P.F. 1972a *Notes on the chemical and biological analysis of Irish waters.* Dublin. An Foras Forbartha

FLANAGAN, P.J. and TONER, P.F. 1972b *The national survey of Irish rivers. A report on water quality.* Dublin. An Foras Forbartha.

HAWKES, H.A. 1962 Biological aspects of river pollution. In L. Klein (ed.), *River pollution II. Causes and effects,* 311-432. London. Butterworths.

HAWKES, H.A. 1979 Invertebrates as indicators of river water quality (UK). In A. James and L.M. Evison (eds), *Biological indicators of water quality,* 2.1 - 2.45. Proceedings of a symposium held at University of Newcastle-upon-Tyne, 12-15 September 1978. Chichester. Wiley.

HELLAWELL, J.M. 1986 *Biological indicators of freshwater pollution and environmental management.* London and New York. Elsevier.

HYNES, H.B.N. 1959 The use of invertebrates as indicators of river pollution. *Proceedings of the Linnean Society of London* 2, 165-9.

HYNES, H.B.N. 1960 *The biology of polluted waters.* Liverpool University Press.

HYNES, H.B.N. 1970 *The ecology of running waters.* Liverpool University Press.

LUCEY, J. 1987a Slime growth in a river receiving beet-sugar waste. *Irish Journal of Environmental Science* 4, 42-5.

LUCEY, J. 1987b Biological monitoring of rivers and streams using macroinvertebrates. In D.H.S. Richardson (ed.), *Biological indicators of pollution*, 63-75. Dublin. Royal Irish Academy.

LUCEY, J. and McGARRIGLE, M.L. 1987 The distribution of the crayfish *Austropotamobius pallipes* (Lereboullet) in Ireland. *Irish Fisheries Investigations* A, Number 29, 1-13.

MASON, C.F. 1981 *Biology of freshwater pollution.* London and New York. Longman.

McCARTHY, T.K. 1986 Biogeographical aspects of Ireland's invertebrate fauna. In D.P. Sleeman, R.J. Devoy and P.C. Woodman (eds), *Proceedings of the postglacial colonisation conference*, 67-81. Occasional Publication of the Irish Biogeographical Society, Number 1.

McGARRIGLE, M.L. and LUCEY, J. 1983 Biological monitoring in freshwaters. *Irish Journal of Environmental Science* 2, 1-18.

O'DONNELL, C. 1980 *Organic micropollutants in Irish waters. Results of a pilot study.* Dublin. An Foras Forbartha.

REID, D.M. 1939 On the occurrence of *Gammarus duebeni* (Lillj.) (Crustacea, Amphipoda) in Ireland. *Proceedings of the Royal Irish Academy* 45B, 207-14.

REYNOLDS, J.V. 1986 Insect populations in a river receiving acid mine drainage. *Irish Journal of Environmental Science* 4, 35-41.

REYNOLDSON, T.B. 1978 A key to the British species of freshwater triclads. *Freshwater Biological Association Scientific Publications* 23, 1-31.

STRANGE, C.D. and GLASS, G.B. 1979 The distribution of freshwater gammarids in Northern Ireland. *Proceedings of the Royal Irish Academy* 79B, 145-153.

TONER, P.F., CLABBY, K.J., BOWMAN, J.J. and McGARRIGLE, M.L. 1986 *Water quality in Ireland. The current position. Part one: General assessment.* Dublin. An Foras Forbartha.

WARNICK, S.L. and BELL, H.L. 1969 The acute toxicity of some heavy metals to different species of aquatic insects. *Journal of the Water Pollution Control Federation* 41, 280-4.

WARREN, C.E. 1971 *Biology and water pollution control.* Philadelphia. W.B. Saunders.

WATER POLLUTION ADVISORY COUNCIL 1983 *A review of water pollution in Ireland.* Dublin. Water Pollution Advisory Council.

WEBER, C.I. (ed.) 1973 *Biological field and laboratory methods for measuring the quality of surface waters and effluents.* Environmental Protection Agency, Washington, EPA 670/4. 73. 001.

WESTLAKE, D.F. and EDWARDS, R.W. 1957 Director's Report. *Report of the Freshwater Biological Association* 25, 35-7.

WHITTON, B.A. and SAY, P.J. 1975 Heavy metals. In B.A. Whitton (ed.), *River ecology*, 268-311. Oxford. Blackwell.

YOUNG, M. and WILLIAMS, J. 1983 The status and conservation of the freshwater pearl mussel (*Margaritifera margaritifera* Linn.) in Great Britain. *Biological Conservation* 25, 35-52.

In: Steer, M.W. (ed.) 1991 *Irish Rivers : Biology and Management*, pp 151-162.
Royal Irish Academy, Dublin.

THE EFFECTS OF HYDRO-ELECTRIC DEVELOPMENT
ON THE SALMON STOCKS IN THE RIVER LEE

Eileen Twomey

Department of the Marine, Fisheries Research Centre, Dublin

ABSTRACT

Following the harnessing of the River Lee for hydro-electric development in 1956/57 there was a complete collapse of salmon stocks
in 1960.

This paper describes investigations into the causes of the decline
in catches which followed the damming of the main river channel. A
deterioration in water quality in the tail-race downstream of Inniscarra Dam caused by the decaying vegetation in the new reservoirs
brought about mortality of adult fish from 1956 onwards. Predation
by pike and trout on the descending smolts in the reservoirs and
damage to smolts in their passage downstream through the turbine
shaft were identified as the principal factors to the loss of salmon in
the system.

Measures taken to remedy the losses of wild breeding salmon
failed. Artificial rearing was undertaken and resulted in a partial
restoration of the salmon run.

INTRODUCTION

The River Lee rises in Gougane Barra lake, Co. Cork, and flows in an
easterly direction to Cork city before discharging into Cork harbour. The
main channel measures 88 km in length and the tributary streams have a
total length of 176 km. The catchment area is 124,000 hectares (Fig. 1).

The harnessing of the River Lee for hydro-electric development was completed in 1957. It required the creation of two dams, one at Carrigadrohid
and the second 6.25 km downstream at Inniscarra. The head at Carrigadrohid is *c.* 14 m high and at Inniscarra *c.* 30 m high.

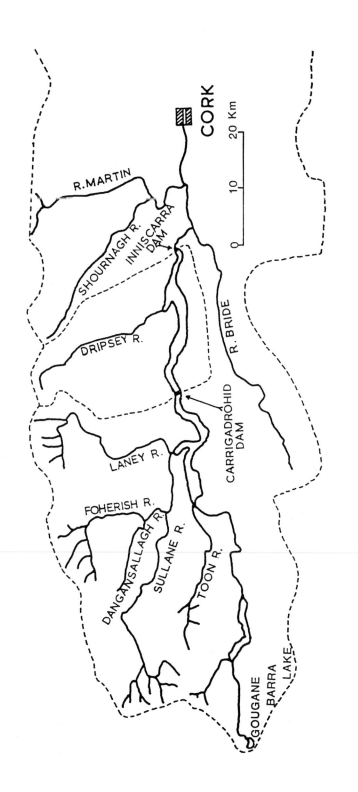

Fig. 1. River Lee catchment showing the major tributary streams.

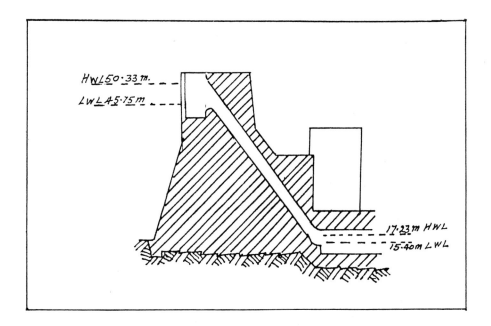

Fig. 2. Section through the fish pass.

Adults pass upstream through a Borland fish pass which was constructed at the upstream face of both dams. A detailed description of this type of fish pass is given in O Meallain (1951). It consists of a shaft extending from the tail-race to the head-water level at the power station (Fig. 2). Water flows through the pass and it attracts fish into the lower chamber at tail-race level. The lower gate closes and the fish are lifted by the water level to the top of the dam and released into the head-waters at reservoir level. Smolts can pass downstream (a) through the turbine shafts, (b) over the spillway and (c) through the fish pass.

The salmon stocks declined dramatically after the dams were built. This paper examines the causes of the decline and describes the subsequent partial rehabilitation of the fishery.

MATERIALS AND METHOD

The mortality of adult fish in the tail-race downstream of Inniscarra Dam was investigated in relation to the water quality in the reservoir and tail-race.

Electrofishing was carried out in February 1962 to estimate the population of juvenile salmon in the tributary streams flowing into both Carrigadrohid and Inniscarra reservoirs.

Pike were removed from the reservoirs by gill netting and trapping. All pike taken between February and June in 1962 and 1963 were weighed and measured, and stomach contents in a proportion of the fish were examined. The food was classified into smolts, other fish, amphibians and invertebrates.

Between 1962 and 1969 there was an upsurge in the population of trout. Trout were sampled by gill netting during the smolt run of 1969 and their stomach contents were analysed.

RESULTS

Salmon numbers

The numbers of salmon caught in the Lee are shown in Fig. 3 and the numbers counted at Inniscarra in Fig. 4. The average annual catch of salmon for the 10-year period 1949 to 1958 was 8,497. The decline was first noted in 1960 and in the next 10 years (1960-69) the average catch was 753. An electronic counter was installed in the Dorland fish pass in 1957. The count of salmon from 1960 onwards showed a similar decline, reaching an all-time low in 1972 with only 37 fish counted upstream.

Adult fish mortality

Between 1956 and 1958 large numbers of dead and dying fish were found downstream of Inniscarra Dam. It was difficult to get accurate figures as fish were being removed by legal and illegal means in large numbers. The fishery authorities in the area turned a blind eye to the infringement of the law as the salmon were not in a fit condition to migrate upstream. In 1957 it was estimated that more than 2,000 dead or dying fish were removed from the tail-race. There was a decrease in the numbers of dead fish removed in 1958 and 1959. In 1960 virtually no salmon dead or alive were seen in the vicinity of the tail-race.

The flooding of the river channel upstream of the dams resulted in the submersion of the vegetation which on decay caused a pronounced dissolved oxygen deficiency in the hypolimnial layers and to a lesser extent in the metalimnion. The oxygen levels fell as low as 22% saturation during the post-flooding years (Electricity Supply Board, personal communication). The deoxygenated water was discharged through the turbines and through the scour valve at the base of the Inniscarra dam wall. It was evident from the behaviour of the fish that oxygen depletion was the cause of distress. This problem was alleviated by the discharge of water over the spillway at intervals. The water from the surface of the reservoir was rich in oxygen.

Juvenile stocks

In February 1962 approximately one-third of the smolt-producing waters of the Lee were electrofished using four AC generators and one pulsed generator. In this survey salmon parr of less than 100 mm were not counted and it was known that the size range would not smoltify in 1962. Two-thirds of the juveniles counted were between 101 mm and 119 mm and the remainder between 120 mm and 139 mm. The size of the migrating smolts (examined in April and May) was between 100 mm and 159 mm. The majority in April were between 140 mm and 159 mm and in May between 120 mm and 139 mm. The cut-off minimum point of 100 mm was considered reasonable in estimating the numbers of smolts in the system.

The total production of smolts in the system was estimated as 117,200. This figure was far short of the potential from 1959 spawning stock from which the 1962 smolt population was derived. Based on the data available from other systems the smolt production should have been in the region of 400,000. The shortfall in spawning stock could have been due to the delay in the movement of adult fish into the spawning ground. No information was available on the numbers of fish that passed upstream of Carrigadrohid Dam in 1959, but data for 1957 indicated that the escapement into the main spawning grounds was less than 50% of the total escapement. It was assumed that fish destined for the main spawning grounds were unduly delayed and that they shed their eggs in the reservoirs. The flooding of the nursery and spawning area in the main river channel also reduced the smolt production. It was estimated that the loss in area of spawning ground was 35% (C. McGrath, personal communication).

Pike predation

Following the inundation of the Lee Valley, a still water of 930 ha was created upstream of Carrigadrohid Dam and a further, deeper still water of 530 ha was formed at Inniscarra. With the flooding of the hinterland ideal spawning and nursery conditions were created for pike, especially at Carriga-drohid reservoir which was shallow. An estimate of the pike population was made using published data from Carconneely Lake (Anon. 1965) and the higher estimates from Carlander (1955). The figure obtained (102,625 pike) was considered to be an underestimate since between 1956 and 1964 a total of 70,765 pike were removed from the two reservoirs.

An analysis of the stomach contents of the gill-netted pike was carried out. The trap-caught pike were excluded as there was evidence that smolts were also captured in traps. The smolts were therefore a captured prey for the pike.

The predation by pike on smolts is given hereunder on a monthly basis from March to May 1962 and from February to June 1963.

| | 1962 | | 1963 | |
Month	Number examined	% with smolts	Number examined	% with smolts
February	-	-	134	1.5
March	327	3.0	960	1.1
April	179	10.0	599	6.8
May	184	13.0	92	11.9
June	-	-	30	10.0

There were very few smolts in the reservoirs in February and March and this could account for the low predation rate. The percentage of pike eating smolts was highest in April and May and months in which the major down-

156

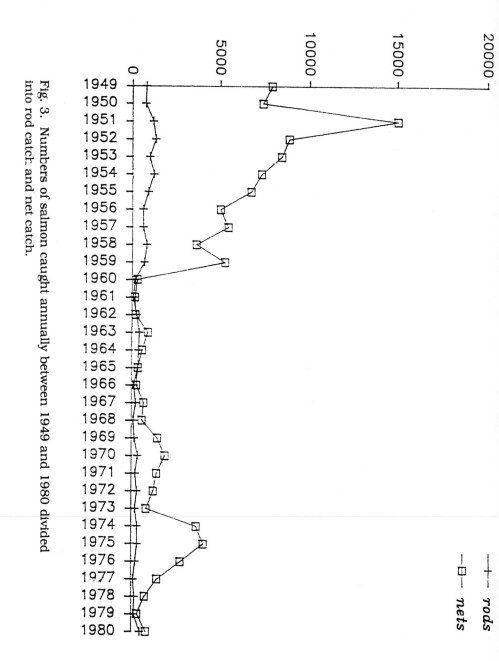

Fig. 3. Numbers of salmon caught annually between 1949 and 1980 divided into rod catch and net catch.

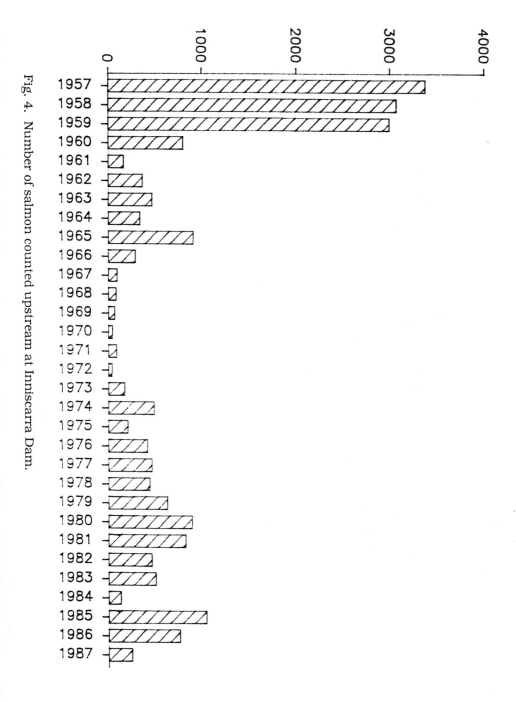

Fig. 4. Number of salmon counted upstream at Inniscarra Dam.

stream migration occurred. Smolts were present in the reservoir for over 60 days. The average consumption of the pike examined was 2 smolts each. Based on a population of 100,000, the population of pike could devour at a minimum 20,000 smolts, one-fifth of the total estimated population of smolts. Smolts were present in the reservoirs over a long period. It can be assumed that pike predation was a serious problem to migrating smolts up to and including 1963.

It was evident from the catch returns, however, that a reduction in the pike population was achieved from 1965 onwards arising from intensive predator control. Another factor which was instrumental in reducing the pike population was the lowering of the reservoirs for maintenance of the power stations. The drawdown of water in the reservoirs from February onwards to cater for maintenance succeeded in achieving a substantial reduction in the pike population. Pike spawned in the flooded margins of the reservoirs, and when these margins were left high and dry the eggs and young fry failed to survive owing to dessication of their spawning and nursery grounds.

Predation by trout on smolts

With a reduction in the pike population there was an increase in the trout population in the reservoirs. Very favourable conditions were created for trout in terms of food supply. A similar situation arose in the 1940s in Poulaphouca as a result of the hydro scheme on the Liffey.

To compensate for the decline in salmon runs the Electricity Supply Board stocked the tributary streams with large quantities of unfed fry and fingerlings from 1964 to 1970. Electrofishing of the tributary streams between 1965 and 1969 indicated a good survival of the stocked fish and smolts were seen in the reservoirs following the stocking.

Investigations into predation by trout were carried out in 1969 during the smolt run. Netting for trout was carried out at different locations in the system. As can be seen from Table 1, eleven different areas were selected for netting and smolts were found in the stomach contents of a proportion of the trout taken, except at Coolcower and Dripsey. The highest predation was found in trout taken in front of the dam wall in the Carrigadrohid area, where 115 trout were examined and 79 (68.6%) were found to contain an average of two smolts each (one trout of 1 kg had eight smolts in its stomach). At Inniscarra Dam wall similar results were obtained. The overall picture of predation by trout can be summarised by stating that of the feeding trout examined (317) 48.2% were feeding on smolts. By 1969 the trout had superseded the pike as predators. The wild run of fish had virtually disappeared by 1964. The stocked fry and fingerlings provided the fodder fish for the trout.

Passage of smolts through the fish pass and turbine intake

There were three exits available for the passage of smolts downstream of the dams. An outlet was provided at the fish pass and smolts could also migrate downstream through the turbine shaft. Smolts are also capable of descending over the spillways.

From observations at the dam walls it was evident that smolts were experiencing delay in their descent. Smolts were seen milling around in front of the dam wall but made no effort to enter the pass in any numbers. Various devices were installed upstream of the fish pass to attract smolts but these were not successful. Only when the level of water in the the fish pass was lowered to 6" approximately below the level of the reservoir were the right conditions created to attract smolts into the fish pass. But the station operation made it impossible to implement this regime. In 1963, between April 23 and May 10, a total of 7,333 smolts passed downstream but this was only a small proportion of those observed upstream of the dam wall. A trap consisting of a conical-type net was fitted into the base of the 4 megawatt turbine with a live box attached on the downstream end. From April 10 to June 13 a total of 298 smolts were taken in a dead or dying condition and 30 were released alive. But as most of the live fish were descaled on release their chance of survival was poor. The dead fish collected had obvious mechanical injuries including decapitation and external gashes etc. Injuries due to pressure were also noted: haemorrhage of the eye or head region, diffused blood in the muscles and burst swim bladders.

A search of the tail-race was also carried out. Owing to poor visibility and depth of water this proved a difficult task. But in one afternoon when the turbines were off load, approximately 103 smolts were picked up and also eel pieces. All of these fish had injuries indicative of turbine damage. It was evident from these investigations that turbine damage was a contributory factor in the reduction of fish stocks in the Lee. No investigations were possible in relation to the 15 megawatt turbine at Inniscarra or the 8 megawatt turbine in Carrigadrohid as the force of water was too strong to attach a net to the base.

Rehabilitation measures for the restoration of the stocks

It was evident from the investigations carried out between 1957 and 1969 that the damming of the River Lee for hydroelectric production resulted in a complete collapse of the salmon stocks. In order to compensate for the loss the Electricity Supply Board constructed a smolt rearing station for the production of 150,000 smolts. Stocking with hatchery-reared smolts was carried out downstream of Inniscarra Dam. The waters upstream were developed as trout and coarse fisheries.

A number of release techniques based on size of smolts, time and place of release were carried out. The results of these experiments were monitored using the microtagging technique described by Browne (1982). The results obtained showed that highest recoveries of adults derived from hatchery-reared smolts were obtained when the smolts were released between April and early May in the lower reaches of the river. The recoveries varied from 0.55% for a batch released in early February immediately downstream of Inniscarra Dam to a recovery of 2.75% for a batch released in the lower freshwater reaches in early April. A batch released in the tidal waters in

April gave a recovery rate of 2.35% (J. Browne, personal communication).

The hatchery-reared smolts are being exploited right around the coast from Donegal to Waterford. Proportionately more smolts of Lee origin were being exploited in the high seas fisheries in West Greenland than from any other river in Ireland. Despite the exploitation without the catchment there has been an increase in the estuarine nets and in the rod fishery. The sampling of fish taken for hatchery purposes has shown a steady increase in the proportion of hatchery-reared fish in the main channel when compared with the wild component. In 1974 approximately 50% of the stock were of hatchery origin. The proportion has increased to 98% to 100% in recent years (N. Roycroft, personal communication).

DISCUSSION AND CONCLUSIONS

The investigations carried out between 1957 and 1969 showed that the hydro-electric scheme on the River Lee was responsible for the collapse of the stocks from 1960 onwards.

The decaying vegetation caused deoxygenation of the hypolimnion layer of Inniscarra reservoir. The discharge of deoxygenated water through the turbines and the scour valve at the base of Inniscarra Dam caused mortality of adult fish. A study in the Norris reservoir, USA, showed that a pronounced dissolved oxygen deficiency characterised the hypolimnial waters during the first three years after filling (Anon. 1969). To counteract the effects, alternate discharge regimes of epilimnial (high level) and hypolimnial (low level) were recommended. The discharge of water over the spillway at Inniscarra increased the oxygen level in the tail-race and resulted in a decrease in the mortality of adult salmon.

The count of fish at Inniscarra Dam indicated that adults passed upstream without difficulty in the years 1956 to 1959. The problem arose with the descent of smolts. Investigations carried out showed that the predation on smolts by both pike and trout caused high mortality. Mills (1964) estimated that pike consumed 10% of the total smolt run in the River Bran in Scotland which was obstructed by three hydro-electric dams. The percentage of pike with smolts in their stomachs in the Bran system varied from 37.5% in the uppermost reservoir to 9.3% in the lowest reservoir. Smolts occurred in pike stomachs in April and May and occasionally in June. In the River Lee, pike predation varied between 13.0% and 6.8% in the pike sampled from April to June. Predation by trout on smolts was also high in the Scottish reservoirs. In April and May 35.3% of the trout examined contained smolts. In the River Lee over a similar period 36% of the trout examined contained smolts.

Smolts tended to congregate in large numbers in front of the dam wall, failing to find the entrance to the fish pass. Ruggles (1980) suggests that smolts moving downstream in rivers with high velocity tend to avoid pike as pike prefer slower velocities. This could account for the high predation rate as well as the delay experienced at the Dam.

The mortality of smolts in the tail-race was attributed to turbine damage as the majority of the smolts examined were found to have external injuries, abrasions or descaling. It was not possible to estimate the number of smolts using the turbine shafts in preference to the fish pass exit. Ruggles (1980) reviewed the literature and identified hydro-electric turbines as a source of injury to downstream migrating fish.

It is mandatory under the Fisheries (Consolidation) Act 1959 (Anon. 1959) to provide screens at the inlet to the turbines in the case of mill dams to prevent the entry of smolts during the seaward migration. Smolt screens are put in place between March and June each year. The Electricity Supply Board are exempt from compliance as it would not be practical to maintain screens at the depth of the intakes in most of the large hydro dams in this country. Small-scale hydro dams are compelled to provide screening.

Another possible source of mortality was the passage of smolts over the spillway. Schoeneman et al. (1961), cited by Ruggles (1980), estimated mortality of Chinook salmon yearlings passing over the spillway (27 m head) on the Columbia River at 2%. The head at Inniscarra is 30 m but the relative importance of the spillway in smolt mortality was not investigated.

The restoration of the nursery areas of the River Lee by stocking with juvenile salmon was successful in producing smolts but did not succeed in restoring the runs of adult fish because of the high smolt mortality owing to various factors. The decision taken in 1970 to construct a smolt-rearing station succeeded in restoring in part the salmon runs into the River Lee. It was not possible to restore the very valuable early run fish into the Sullane and its tributary, the Foherish. These early run two sea water fish were the main component of the Lee stocks (Newman 1958).

The development of the reservoirs for coarse fishing and to a lesser extent trout fishing has proved a valuble asset in an area where still-water fishing is scarce.

ACKNOWLEDGEMENTS

I wish to express my sincere thanks to Noel Hackett of the Central Fisheries Board who coordinated the collection of pike in the reservoirs and did the analysis of the stomachs; to my colleague R.D. Fluskey who analysed the stomach contents of the trout; to my engineering colleagues Charles McGrath and Donal Murphy who cooperated in designing the experiments at the turbine shaft and the collection of fish in the tail-race; the Electricity Supply Board staff both at the power stations and headquarters who were always most helpful; to my late collegue E.D. Toner who supervised and guided this study until his untimely death in 1965; to my colleague Christopher Moriarty who read and made constructive comments on the manuscript; and to Anne McDaid who typed the manuscript.

REFERENCES

ANON. 1959 Fisheries (Consolidation) Act 1959. Dublin. Stationery Office
ANON. 1965 Inland Fisheries Bulletin. Dublin. Stationery Office
ANON. 1969 S.F.I. Bulletin, no. 210. Sport Fishing Institute, Washington D.C. 20005.

162

BROWNE, J. 1982 *First results from a new method of tagging salmon - the coded wire tag.* Fishery Leaflet No. 114. Trade and Information section, Dept of the Marine, Dublin 2.

CALDERWOOD, W.L. 1945 Passage of smolts through turbines. Experiments at a power station. *Salmon Trout Mag.* 115, 214-21.

CARLANDER, K.D. 1955 Standing crop of fish in lakes. *J. Fish Res. Biol. Canada* 2, 543-70.

MILLS, D.H. 1964 The ecology of the young stages of the atlantic salmon in the River Bran, Ross-Shire. *Freshwater and Salmon Fisheries Research*, no. 32. Dept. Ag. Fish, Scotland.

NEWMAN, H. 1958 Salmon of the River Lee 1944 and 1945. *Proceedings of the Royal Irish Academy* 59B, 53-69.

O MEALLAIN, S. 1951 Fish pass at Leixlip Dam. *Journal of the Department of Agriculture* 48, 3-10. Dublin. Stationery Office.

RUGGLES, P. 1980 A review of the downstream migration of Atlantic salmon. *Canadian Technical Report of the Fish and Aquatic Society* no. 952.

SCHOENEMAN, D.E., PLESSEY, R.T. and JUNGE, C.O. 1961 Mortalities of downstream migrant salmon at McNary Dam. *Transactions of the American Fisheries Society* 90, 58-72.

In: Steer, M.W. (ed.) 1991 *Irish Rivers : Biology and Management,* pp 163-184. Royal Irish Academy, Dublin.

DETECTING AND MANAGING THE INFLUENCE OF FORESTRY ON RIVER SYSTEMS IN WALES: RESULTS FROM SURVEYS, EXPERIMENTS AND MODELS

Stephen Ormerod, Graham Rutt and Neill Weatherley

Catchment Research Group, Department of Pure and Applied Biology, University of Wales College of Cardiff

and

Ken Wade

Welsh Water, Bridgend, West Glamorgan

INTRODUCTION

In a seminal paper, Hynes (1975) recognised the critical linkage which exists between streams and their catchments: features of watersheds such as geology, soils and vegetation could have influences on the physico-chemistry and ecology of the streams and rivers which drained them. Almost in parallel during the 1970s, experiments involving whole catchments at the Hubbard Brook Forest, New Hampshire, reinforced the existence of such links. Catchment manipulation, whether experimental or operational, led to marked changes in the biogeochemistry and ecology of the drainage system (Likens *et al.* 1977). From these and many subsequent studies, the clear inference was that major changes in catchments would be accompanied by changes in lakes and streams which received run-off.

Many activities by man qualify as having major impacts on surface waters in this way. In the United Kingdom, plantation forestry in particular has come under increasing scrutiny. In Wales, around 11% of the land area is now occupied by forest, but this is concentrated in upland areas over 200 m, where around 25% of land is covered (e.g. Parry and Sinclair 1985). Many Welsh rivers therefore either rise in forest or receive drainage from substantial planted areas, a feature of great importance because of the marked change in catchment attributes which result. As a consequence, there have been several reviews of the effects of forestry in Wales and Scotland, together with specific studies where an influence has been suspected (e.g. Harriman and Morrison 1982; Stoner *et al.* 1984; Egglishaw 1986; Ormerod, Mawle *et al.* 1987). Options for managing some of these effects have also been suggested, and in some cases tested (Brown 1990).

Forestry is now becoming an important issue in Ireland. Around 5% of the land area of the Republic is given to conifer plantation, with 12,000 ha being planted annually (Dr Mary Kelly-Quinn, pers. comm.). This is approximately twice the maximum planting rate ever achieved in Wales (see Ormerod and Edwards 1985), and if maintained would double the area of productive forest in Ireland in 30 years. Concerns have been expressed over influences on the ecology of Ireland's rivers, not least their important salmonid fisheries. It is therefore timely to review critically the ecological effects of forestry as they have been experienced in Welsh rivers. While separate studies will be necessary under the different ecological conditions which prevail in Ireland, at least some of the impacts and processes noted in Wales will be relevant to the Irish situation.

This review outlines evidence on the effects of forestry in Wales, examining in detail a case study on the postulated role of conifers in exacerbating surface water acidification. It describes progress in understanding the processes involved, and outlines the potential for management.

POSSIBLE PATHWAYS OF FOREST INFLUENCE

The possible effects of forestry operations on the ecology of surface waters have been summarised in Fig. 1. These effects could occur through physical, energetic and chemical processes. Through these pathways, there could be direct or indirect ramifications for the ecology of organisms in all trophic levels, with consequences for fisheries and conservation. In addition, there might be effects on economic uses of water for such purposes as abstraction and power generation, though these are not the subject of this review.

The exact blend and intensity of effects will be determined by many features, including the attributes of the forest and its location, and the stage of forest rotation. Some impacts could arise, for example, during initial ground preparation, whilst others will be delayed until felling. Ameliorative measures will also be important in preventing effects, though the application of suitable management strategies depends on:

1. recognising that an effect will occur, or is occurring, as a result of forestry operations;

2. identifying the target which requires protection (e.g. salmonid nursery, rare species A, breeding bird species B);

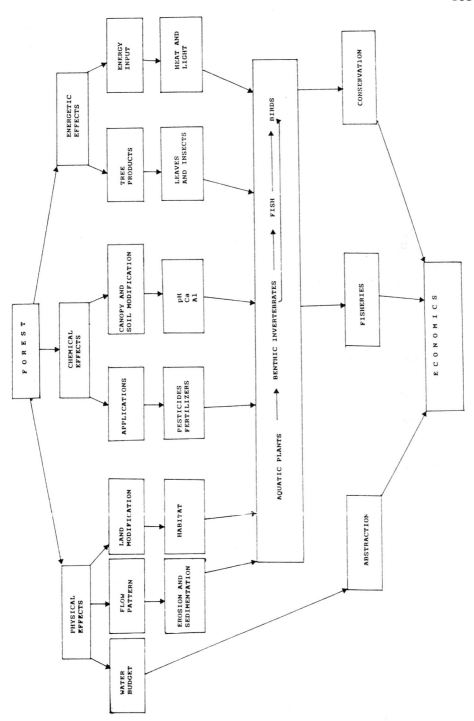

Fig. 1. Selected pathways of influence by forest on aquatic fauna and water utilisation.

3. identifying the processes and pathways involved so that management action can be directed appropriately;

4. having suitable technology to implement management action.

Ideally, these stages would involve a justifiable, modelling framework, so that action could be directed in a cost-effective way.

In many cases, however, the suggestion of ecological effects by forestry is based on extrapolation. Even in instances where physico-chemical influences have been well documented, reviewers have surmised the ecological consequences from knowledge of how such changes sometimes affect organisms in other surface waters, usually unconnected with forests (e.g. Egglishaw 1986; Ormerod, Mawle et al. 1987). This applies to many of the postulated effects including flow changes, erosion, turbidity and sedimentation, alterations in the nutrient status of rivers, decreased light penetration and alteration in the quality, quantity and utilisation of allochthonous inputs (e.g. deciduous litter v. conifer litter). Real field data on the biological effects of these features of forest streams are clearly overdue.

One exception from Wales has been a recent data set which showed differences in the habitat structure of forested and moorland streams, probably because rapid drainage and shading often lead to erosion of the stream margins (Rutt et al. 1989) (Fig. 2). Invertebrates usually show distinct habitat requirements, and some of the species scarce in forest streams are those dependent on marginal habitats (Rutt et al. 1989; Ormerod, Weatherley and Merret 1990). These effects are apparent both in acidic and circumneutral streams (Rutt, unpubl.). Water temperature may also be modified by afforestation. Extensive measurements have recently been made in streams around Llyn Brianne, mid-Wales, and likely effects on stream biology have been assessed by modelling (Weatherley and Ormerod 1990). The procedure has been to use real temperature data from streams draining land of contrasting use to drive equations which predict the growth or development of organisms. These modelling studies show, for example, that trout growth could be impaired in forest streams by comparison with moorland (Fig. 3a). Invertebrate growth could also be retarded to the extent that temperature conditions in forest streams could exclude facultative summer generations in some species (Fig. 3b). There are some potential problems in this approach in that the growth equations are derived from laboratory studies, few of which mimic accurately the dynamic nature of temperature changes in real streams. Field data would again be useful.

A CASE STUDY: ACIDIFICATION AND ITS ECOLOGICAL EFFECTS

The area of forestry that has received the greatest attention recently is the possible exacerbation of acidification of surface waters draining from plantations on base-poor soils (e.g. Stoner et al. 1984). This forms a useful case study of the stages of recognising an effect by forestry, and developing a management strategy.

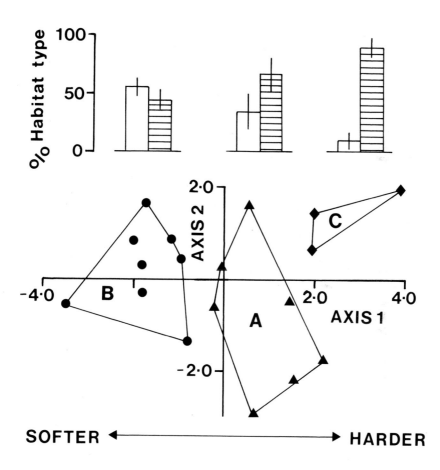

Fig. 2. An ordination of habitat features in streams in the upper Tywi in mid-Wales using principal components analysis. Group A are forest streams, group B acidic moorland streams and group C circumneutral streams which have a steep gradient. Axis 1 can be considered a gradient from 'softer' to 'harder' margins, and the histograms (± S.E.) show the percentage contribution by 'soft' features (e.g. vegetation) and 'hard' features (shaded bars; soil, stones and rock) to the habitat characteristics in each group. Invertebrate taxa associated particularly with group B include *Cordulegaster boltonii, Pyrrhosoma nymphula, Leptophlebia marginata, Halesus radiatus, Elaeophila* sp.

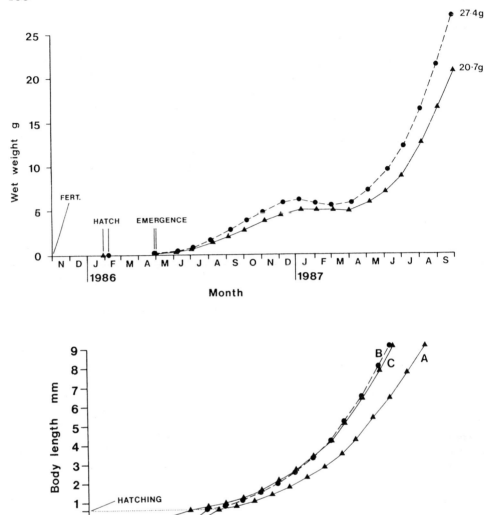

Fig. 3. (a) The simulated development of brown trout from fertilisation to the second year of life calculated from temperature in moorland (●) and forest (▲) streams at Llyn Brianne. (b) The simulated development of *Baetis rhodani* nymphs in moorland (●) and forest (▲) streams calculated from temperature. An egg laid in early May in a forest stream (A) would be unable to complete development to emergence within the summer period, but could in a moorland stream (B). Only eggs laid in early March could complete a summer generation in a forest stream (C), and would require an adult to have emerged by this time. For details of the development equations, see Elliott (1975), Elliott *et al.* (1988) and Crisp (1988).

Recognition of effects

Initial assessments of the acidifying role of forestry came from comparisons between adjacent moorland and forest streams in Scotland (Harriman and Morrison 1982), Wales (Stoner *et al.* 1984; Ormerod and Edwards 1985; Reynolds *et al.* 1986) and northern England (Bull and Hall 1986). In each of these cases, streams draining forest generally had elevated concentrations of aluminium, and sometimes a mean pH lower by 0.5 - 1.0 pH units than streams in moorland. Differences are particularly pronounced at high flow, when pH in afforested streams can fall to pH 4, and aluminium can increase to over 1 mg l^{-1}. Welsh lakes draining different land use also show differences in mean pH and aluminium concentration (Stoner and Gee 1985). Only soft-water areas (<15 mg $CaCO_3$ l^{-1}) have been considered as sensitive to such effects because similar patterns have not been detected in systems where run-off is harder (e.g. Jones 1986). The phenomenon has also been described principally in areas with acidic deposition (e.g. Wells *et al.* 1986; Wells and Harriman 1987), including Wales. However, an increase in the magnitude or frequency of acid episodes is also theoretically possible in areas where sea salts dominate rainfall chemistry (Sullivan *et al.* 1988).

Recognising targets

Biological effects associated with acidification in forest streams include fish populations which are sparse or absent, the absence of some invertebrate groups, and changes in the quality of primary producers (Stoner *et al.* 1984) (Table 1). Some experimental evidence supports the role of increased aluminium concentrations in being directly responsible for fish mortality and disturbance to some invertebrates (Stoner *et al.* 1984; Ormerod, Boole *et al.* 1987), although indirect effects through altered trophic structure may also be involved (Ormerod, Wade *et al.* 1987).

Quantitative differences between the biology of moorland or forest streams, such as in invertebrate density biomass or primary production, have been less obvious than qualitative differences (Harriman and Morrison 1982; Weatherley *et al.* 1989; Ormerod and Wade 1990). Nevertheless, knock-on effects to organisms at higher trophic levels have been noted. The riverine dipper, *Cinclus cinclus*, a bird closely dependent on benthic invertebrates as a food source, is scarce along acidic streams in Wales and Scotland, particularly those draining forest (Ormerod and Tyler 1987; Vickery 1989). Here, the effect appears to reflect the absence of mayfly nymphs and some caddis larvae under acid conditions, because these two groups are important to the birds when feeding nestlings. Reduced supplies of calcium at low pH might also be involved, however (Ormerod, Bull *et al.* 1988).

The biological targets of forest effects through acidification are thus ecosystem processes such as energy transfer, ecosystem quality in several trophic levels, fisheries, and features of prime conservation importance such as birds.

Table 1. Some biological characteristics of streams around Llyn Brianne. (All concentrations (except pH) are in mg l^{-1}.)

Stream type	pH	Ca	Al	Diatom	Characteristic biota: Macroinvertebrate	Fish m^{-2}
Circumneutral moorland	6.9	2.5-5.0	0.04-0.07	*Acnanthes minutissima*	*Baetis rhodani*	0.12-2.79
Circumneutral, deciduous woodland	6.3	1.75	0.05	Not sampled	Stonefly 'shredders'	N/A
Acidic moorland	5.0-5.9	0.8-1.2	0.07-0.16	*Eunotia tenella*	Stonefly 'shredders', *Leptophlebia marginata*, *Isoperla grammatica*	0.0-9.1
Acidic, conifer forest	4.6-5.3	1.05-1.80	0.17-0.48	*Eunotia vanheurkii*	Stonefly 'shredders'	Absent

Note: Only one deciduous woodland stream was included, which is impassable to fish.

Critical assessment of the evidence

One major criticism which might be presented against the evidence for an effect by forestry on water quality is that nearly all of the studies quoted have involved comparisons in space between afforested and moorland catchments. Although all the catchments involved are adjacent (e.g Llyn Brianne/ Plynlimon, Loch Chon/Duchray, Esk/Duddon), such comparisons do not preclude confounding differences which might have occurred even before afforestation (e.g. soil type or geology). This would be especially problematic if forest had been planted preferentially on certain soil types. As a result, some foresters have been prompted to suggest that only long-term experiments would be suitable to assess the real extent of any forest effect (e.g. Bancroft 1988).

Historical data on water chemistry in upland areas are generally insufficient to enable changes following afforestation to be assessed from past records (Warren, Bache *et al.* 1986). Even where data are available, they are characterised by gaps in the record, or by availability over different time-scales (e.g. Ormerod and Edwards 1985). At present, only one study has followed forest development from the initial ploughing stage (Llanbrynmair Moors, Powys), and has not shown any early evidence of acidification or aluminium mobilisation (Warren, Alexander *et al.* 1988). This might be expected if forestry had effects only during subsequent growth (see below), though at this site a mire may have important influences on stream chemistry. Three studies have followed forest development during stages of tree growth, although trends at two have not been evaluated (Loch Ard, Llyn Brianne LI 8) because too few years' data are available to control for the effects of other sources of variability between years (e.g. changing deposition quality, climate). The third site, on the Green Burn at Loch Dee, does not yet show a forest effect (Warren, Alexander *et al.* 1988), though only eight years' data are available. Clearly, there is a need for a larger number of these long-term studies, in view of the importance of this approach, so that the response at such a small number of sites can be assessed (cf. Hurlbert 1984). Additionally, they should cover a longer time-span than those so far available so that all the possible mechanisms of acidification can be evaluated (see below). The lead time for completing experiments of this type is clearly large, and probably decades.

An alternative method of assessing long-term trends in acidification, pioneered in Britain by Battarbee *et al.* (e.g. 1988), has been the use of fossil diatoms in lake sediments. Evidence from the most sensitive sites (<1.5 mg 1^{-1} Ca) in areas of Wales and Scotland subject to acid deposition shows that they began acidifying in the absence of afforestation (Battarbee *et al.* 1988). It has not yet been possible to determine if forestry has any additional effects in these cases, such as increasing aluminium concentration. At less sensitive sites (1.4 - 2.6 mg Ca 1^{-1}), afforestation appears to have had a marked effect on pH, but only some 10-20 years after planting (e.g. Loch Chon v. Loch Tinker ; Battarbee *et al.* 1988). So far, there are no reliable diatom data from lakes in Wales which span the full development of a forest.

Partly because of the lead time involved in obtaining sufficient and representative long-term data, only spatial patterns permit an immediate assessment of relationships between forest presence and stream chemistry. Clearly, the number of catchments in such spatial analysis would have to be large and spanning a range of soil conditions, so that the influence of confounding effects could be minimised. In Wales, data which satisfy these criteria were collected by the Welsh Water Authority in 1984 from 104 catchments on streams throughout 4,000 km^2 of upland Wales (Ormerod, Donald et al. 1989). In three different categories of total hardness, percentage forest cover and aluminium concentration were closely correlated (Table 2), though the slopes of the three regression lines were different (see Fig. 4). There were also marked correlations in this data set between aluminium concentration and several measures of biological status (e.g. Ormerod, Allinson et al. 1986; Ormerod, Bull et al. 1988; Wade et al. 1989). If there had been some selection of certain soil types for forest planting, some correlation between forest cover and hardness might have been expected, but none was found (Table 2). Subsequent analysis of data from Galloway (Harriman et al. 1987) revealed similar trends, though only for streams (Fig. 5).

Identifying the processes

The role of forestry in enhancing aluminium concentration, and/or reducing pH, would be supported further if processes consistent with this effect were identified. Several have been suggested, and were summarised by Stoner and Gee (1985) and Warren, Alexander et al. (1988) to include:

1. removal of base cations by tree uptake;

2. oxidation of sulphur and nitrogen in organically rich soils resulting in the mobilisation of SO_4^{2-} and NO_3^-;

3. alterations in atmospheric inputs, notably mobile anions, by the scavenging of airborne material and by increased evapotranspiration from afforested catchments;

4. alterations in site hydrology, resulting in reduced residence time of water, and reduction in base-flow contribution.

Depletion in base cations. A depletion in base cations in soils around tree roots can sometimes occur (e.g. Billett et al. 1988), but is not likely to explain changes in forest soils such as increased sulphate and aluminium concentration. As Nilsson et al. (1982) have pointed out, the uptake of bases by roots, though involving a balancing output of H$^+$, does not involve the mobilisation of any anion which could transfer acidity to drainage waters.

Oxidation of organic S and N. Many organic soils, including peats, are rich in organic sulphur and nitrogen. When dried by ploughing or by reduced water inputs, oxidation can occur with the result that NO_3^- and SO_4^{2-} arise as oxidation products (Braekke 1981). Whilst this process probably occurs in

Table 2. Correlation coefficients between forest cover and the chemistry of Welsh streams in three different ranges of total hardness. (Data from both Llyn Brianne and Welsh Water Regional Survey, with mean values from October-March incl.)

	Total hardness mg/l:			
	<10	10-15	>15	All
Filterable aluminium (Log)	0.65	0.76	0.61	0.62
pH	-0.41	-0.67	-0.74	-0.36
Total hardness (Log)	0.16	-0.14	0.03	0.04
N of sites	68	23	20	113

(Values underlined are significant at 1%)

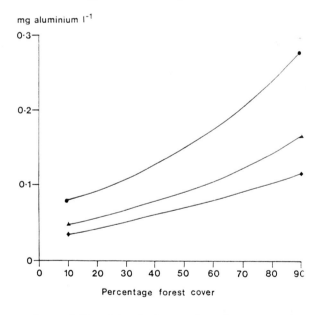

Fig. 4. Regression relationships between forest cover and mean aluminium concentration during winter in Welsh streams of different total hardness (● < 10 mg $CaCO_3$ l^{-1}; ▲ 10-15 mg $CaCO_3$ l^{-1}; ◆ 15-25 mg $CaCO_3$ l^{-1}). Both regressions at <15 mg hardness were significant at P <0.001, and the other at P <0.01, based on F ratios. Statistical parameters were derived on log transformed aluminium. Correlations from the same data are in Table 2.

Table 3.   Rainfall and throughfall chemistry at different sites in Wales over periods of one year. The values are means in mg l⁻¹.

	Llyn Brianne		[1]Plynlimon		[1]Beddgelert	
	pH	SO$_4$	pH	SO$_4$	pH	SO$_4$
Close canopy sitka spruce	4.3	-	4.5	5.1	4.1	3.1
Young spruce	4.3	-	-	-	-	-
Molinia	4.7	-	-	-	-	-
Nardus/ Festuca			4.6	2.5	-	-
Bulk precipitation	4.6	-	4.6	2.4	-	-

[1]   Volume weighted concentrations.

Sources: Llyn Brianne, Gee and Stoner (1988) ; Plynlimon, Reynolds *et al.* (1988) ; Beddgelert, Hornung *et al.* (1987).

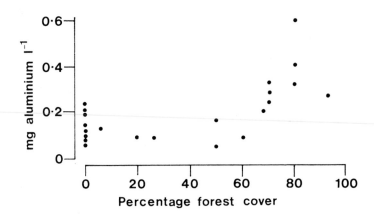

Fig. 5. The relationship between mean aluminium concentration and forest cover in streams in Galloway, south-west Scotland (after Harriman *et al.* 1987). The relationship is best described by the regression, Log aluminium = 2.03 + 0.0044 forest cover, F 1/21 = 8.38, P<0.01, r^2 = 0.29.

afforested areas (possibly following ploughing), its contribution to the flux of sulphate and aluminium has yet to be determined. Comparative studies in areas subject to different acid deposition would clearly prove valuable in making such determinations.

Alterations in atmospheric inputs. Data on the general quality of deposition over Wales have recently been provided by Donald and Stoner (1989), and more recent information is provided by Campbell *et al.* (1989). These provide for comparison with measures of throughfall and stemflow chemistry under-neath the canopies of different tree species at intensively monitored sites such as Plynlimon and Llyn Brianne in mid-Wales, and Beddgelert in the north. At each of these sites, values for SO_4^{2-} and H^+ are considerably ele-vated in throughfall beneath sitka spruce (*Picea sitchensis*) by comparison with bulk precipitation (Table 3). Stemflow concentrations are enriched even further. General reviews suggest that sulphate and chloride inputs may be increased by 50 - 150% by a conifer forest canopy in comparison with open moorland (Warren, Alexander *et al.* 1988). Because determinands such as chloride are relatively conservative, these features have generally been inter-preted as evidence that the conifer canopy is an efficient scavenger of air-borne material. The role of conifers in enhancing evapotranspiration is now well documented (Calder and Newson 1979), and could be important in con-centrating any material deposited onto the conifer canopy either in rainfall or occult deposition.

While there is some debate over whether the tree itself could contribute inorganic solutes to throughfall and stemflow (e.g. SO_4^{2+}, Cl^-), it seems that the effect of a conifer canopy *per se* in Wales is to increase the flux of mobile anions to the forest floor. This effect appears also to increase with tree age, vary between tree species (e.g. larch increases acid deposition more than sitka spruce), and depend on the precipitation chemistry at the site. For example, differences in sulphate concentration between precipitation and throughfall are less marked in remote locations in Scotland (Miller 1984). The scavenging role is supported also by studies at Kershope, in northern England, where aluminium and sulphate concentrations have fallen in streams draining an area felled in 1983, relative to reference streams (J. Adamson, personal communication).

Given the important role of the mobile sulphate ion in surface water aci-dification (Reuss *et al.* 1987), any enhanced input due to the forest canopy clearly would have important corollaries for soil solutions and run-off. In Wales, soil solution samples have been compared between conifer and moor-land canopies in adjacent areas at Beddgelert, Plynlimon and Llyn Brianne. Most solutes show higher concentrations in forest soils by comparison with the same soil types under adjacent moorland, though chloride, sodium, sulp-hate and aluminium show the greatest increase (Hornung *et al.* 1987; Rey-nolds *et al.* 1988; Table 4). Clearly, it is not unreasonable to assume that such enhancement of aluminium concentrations in forest soils is consistent with the increased concentration shown in forest run-off. However, there is still a need to demonstrate quantitatively the effects of enhanced soil alumi-nium on the flux of aluminium in the stream. This is particularly difficult because hydrological pathways through catchments are seldom accurately known (Likens 1984), though further work is in progress on this aspect of forests at Llyn Brianne.

Table 4. The chemistry of soil waters from the B horizon of stagnopodzols under spruce, oak and *Molinia* at Llyn Brianne (Gee and Stoner 1988) and Plynlimon (Reynolds *et al.* 1988). The data are annual means in mg l^{-1}.

Vegetation type	pH	Al	SO$_4$
Llyn Brianne			
25 year spruce	4.3	2.10	8.0
12 year spruce	4.3.	0.78	6.9
Oak	4.5	0.56	5.2
Molinia	4.3	0.81	3.7
Plynlimon			
Spruce	4.3	1.45	9.9
Moorland	4.3	0.39	3.4

Alterations in site hydrology. Proposed alterations in site hydrology by plantation forestry include those due to development of macro-pores by the tree rooting system, and those due to pre-afforestation ploughing and draining (Warren, Alexander *et al.* 1988). The net result in streams draining forest with plough furrows parallel to hill-slopes was to reduce times-to-peak in the hydrograph, and increase peak flows (see Ormerod, Mawle *et al.* 1987). These changes are consistent with more rapid run-off, and hence decreased contact time for buffering reaction. More recent ploughing methods, with contour ploughing and drains prevented from entering the stream, may alleviate hydrological problems, though the techniques are still being evaluated (e.g. Francis and Taylor 1989).

At Beddgelert, forest streams contain 3 - 5 times more aluminium than adjacent moorland streams, even though this site was not ploughed and drained prior to planting. These data have led some authors to suggest that the major acidifying effect of forestry is increased anion capture (Warren, Alexander *et al.* 1988). This is supported by calculations of aluminium sources at Llyn Brianne and Plynlimon, which show that even if all the water in moorland streams were derived from the soil horizons richest in aluminium, the concentrations could still not be as great as those now apparent in forest streams.

Management strategies

Assuming that the above evidence represents a real effect, with conifers exacerbating surface water acidification, the implementation of management action is probably desirable to those concerned with freshwater ecosystems. In general, three kinds of management action are recognisable.

Causative management.   Causative management involves tackling the root cause of surface water acidification, which on the above evidence in Welsh forest streams appears to be enhancement of the effects of acid deposition. Management of the cause would thus involve a reduction in acidifying emissions, and hence reduction in acid deposition. At present the only way to evaluate this strategy is by a modelling approach which allows prediction of the effects of reduced emission. Recent studies at Llyn Brianne have used MAGIC (Model of Acidification of Groundwaters in Catchments) in conjunction with biological models to predict likely biological change under different chemical conditions. These show that even in moorland streams, a reduction in sulphate deposition by 50% of 1984 levels would be required to prevent further acidification and further decline in fish density (Ormerod, Weatherley *et al.* 1988). In forest streams at Llyn Brianne, even greater reductions in deposition would not restore their currently fishless status. These reductions in deposition are substantial when placed in the context of air pollution models. They suggest that even a 90% reduction in UK sulphur emissions would reduce deposition in mid-Wales by only 30-40% (Metcalfe and Derwent 1990). Further cuts in Welsh deposition could thus be achieved only by concerted action across Europe. Moreover, if current understanding of anion capture in the canopy is correct, forest presence could offset the effects of a 25% cut in European emissions, and a 75% cut in UK emissions (Warren, Alexander *et al.* 1988).

Symptomatic management.   The most widely practised method of treating the symptoms of acidification has been the utilisation of lime. In Wales, lakes draining forests have been brought to conditions suitable for fish survival using direct liming (Underwood *et al.* 1987), though the direct liming of streams is more difficult (Weatherley 1988). Application to stream catchments may therefore be necessary in instances where the recovery or maintenance of a fishery is a prime consideration (Warren, Alexander *et al.* 1988), though the wider biological consequences of increasing the pH and calcium concentrations to levels unprecedented in soft-water areas are far from clearly understood (Weatherley 1988). Impacts on terrestrial systems, which hold important conservation resources, will also have to be taken into account.

At Llyn Brianne, stream pH and calcium concentrations have been increased by lime spreading on moorland catchments (Fig. 6a), but forest presence prevents easy application in this way. The bankside clearance and bankside liming of one stream failed to have any lasting effect on stream chemistry, probably because most hydrological flow paths to the stream bypassed the limed area (Brown 1990). In a second case, lime was targeted to

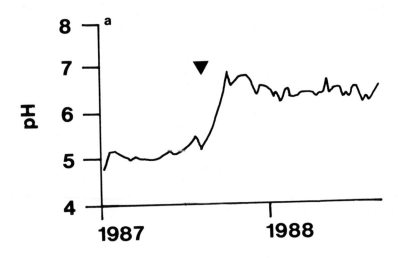

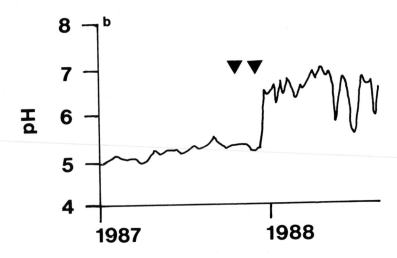

Fig. 6. Changes in the pH of a moorland (a) and forest (b) stream at Llyn Brianne following liming (arrows). In (a) the whole catchment was limed, and in (b) lime was applied to the wetland source at the head of the catchment. More details can be found in Brown (1990).

the hydrological source areas of a forest catchment, which in this case had been left unplanted and accessible. Calcium concentration and pH both increased, but they decline repeatedly at low flow when acid groundwater appears to have an effect on stream chemistry which overrides the effect of 'source' liming (Fig. 6b). The biological consequences of the resulting chemical instability have yet to be evaluated. In a third case, where the hydrological source areas of the stream had been planted and hence were inaccessible, lime was applied by helicopter in a pelletised form to allow passage through the canopy. This treatment proved expensive, and its effectiveness is still being evaluated, though early indications are again of a marked decline in pH at low flow.

So far, there have been no afforestation schemes in Wales where limestone has been incorporated at the planting stage, though several are planned. In these cases, one feature of key importance will be the loss rate through leaching, which will in turn determine the duration before further lime application is required to maintain a suitable pH. The number of such re-applications required during the forest rotation will clearly have a bearing on the economics of timber production.

Preventative management. The difficulties involved in reducing acid deposition or finding a suitable liming strategy have led to the development of preventative management strategies. In Wales, a policy practised by the National Rivers Authority is to identify areas 'at risk' from acidification on the basis of stream total hardness (an indication of buffering capacity), and then to suggest acceptable conifer cover on catchments within the given hardness range (Table 5). According to a survey across the whole Welsh region, these levels of planting are close to those at which mean aluminium concentrations exceed 0.08 - 0.1 mg l^{-1} (Fig. 4). This concentration is often thought of as a threshold for adverse effects on fisheries, and probably also on other organisms (e.g. Ormerod *et al.* 1986).

Table 5. The extent of forest cover suggested in interim guidelines by the National River (Water) Authority in different categories of total hardness.

Mean total hardness (mg CaCO₃ l⁻¹)	Acceptable forest cover
< 12	No planting other than the catchments of first order streams which have minimal effect on the receiving watercourse. Planting should not exceed 10% of the catchment above any point with hardness 15 mg l⁻¹.
12 - 15	Up to 30% subject to agreed planting guidelines.
12 - 25	Up to 70% subject to agreed planting guidelines.

CONCLUSIONS

Clearly, while long-term data are either lacking or contradictory, on balance the conclusion from inter-catchment comparisons is that run-off from forest tends to be richer in aluminium, and sometimes lower in pH, than from moorland. These features are apparent mostly in areas with acid deposition and soft water. Biological differences consistent with low pH and high aluminium are apparent between moorland and forest streams, and there are identifiable processes in afforested systems which could explain their run-off chemistry. If these different pieces of circumstantial evidence did not indicate an effect by forestry, the degree of coincidence in the results would be remarkable. Nevertheless, the effects of some processes require separation (e.g. sulphur oxidation, sulphate scavenging by the canopy, and sulphate cycling by the tree) and there is also a need to demonstrate the quantitative effect of enhanced aluminium concentrations in forest soils on streams which receive drainage. More long-term studies are required.

In managing the effects of acidification in forest streams, there are problems of the scale and expense involved in deposition reduction. There is also no proven method of using lime to neutralise acidity in forest streams with equal effect at all flows, though work is in progress. Loss rates from application in new planting schemes require assessment and the impacts of lime on terrestrial ecosystems requires evaluation. In the meantime, preventative action is being used as a basis for policy by water undertakings in Wales, though this places the goal of protecting freshwaters above the goal of timber production.

The acidification issue forms a useful template for consideration of other physico-chemical effects by forestry since it has progressed to the stage of modelling and management. Ecological studies on these other effects clearly need to extend beyond the extrapolation currently undertaken. As a result of the absence of real field data on these other effects, many of the management strategies sometimes adopted in forest design have yet to be fully appraised in biological terms. Assessment of all the pathways of forest influences will become increasingly pressing if forest streams are manipulated in such a way that their chemistry becomes suitable, for example, for fish survival. Expensive operations aimed at chemical change will be wasted if they only achieve the discovery that forestry imposes biological limits in other ways.

ACKNOWLEDGEMENTS

Our thanks are due to Dr B. Reynolds, Dr Mary Kelly-Quinn, Dr M. Hornung, Dr R.W. Battarbee, Dr J.H. Stoner, Dr A.S. Gee and Dr T. Nisbet for discussing several aspects of this paper. Professor R.W. Edwards made valued comments on the manuscript. The work was largely funded by the Department of the Environment and the Welsh Office, and formed part of the Llyn Brianne project. The views expressed are those of the authors and not necessarily those of the organisations they represent.

REFERENCES

BANCROFT, C. 1988 The Forestry Commission replies. *Rural Wales* (Autumn 1988), 18-20.

BATTARBEE, R.W., ANDERSON, N.J., APPLEBY, P.G. *et al.* 1988 *Lake acidification in the United Kingdom 1800-1986.* London. Ensis.

BILLETT, M.F., FITZPATRICK, E.A. and CRESSER, M.S. 1988 Long-term changes in the acidity of forest soils in north east Scotland. *Soil Use and Management* 4, 102-7.

BRAEKKE, F.H. 1981 Hydro-chemistry of high altitude catchments in southern Norway. 1. Effects of summer droughts with soil-vegetation characteristics. *Report of the Norwegian Forest Research Institute* 36, 1-26.

BROWN, S.J. 1990 The Llyn Brianne Project. In H. Barth (ed.), *The influence of land use on surface water acidification*, 3-17. Proceedings of the COST 612 Workshop, Cardiff, April 1988. Brussels. CEC.

BULL, K.R. and HALL, J.R. 1986 Aluminium in the Rivers Esk and Duddon, Cumbria, and their tributaries. *Environmental Pollution (Series B)* 12, 165-93.

CALDER, I.R. and NEWSON, M.D. 1979 Land use and upland water resources in Britain: a strategic look. *Water Resources Bulletin (U.S.)* 15, 1628-39.

CAMPBELL, G.W., COCKSEDGE, J.L., COSTER, S.M. *et al.* 1989 *Acid rain in the United Kingdom: spatial distributions and seasonal variations in 1986.* DTI, Warren Spring Laboratory.

CRISP, D.T. 1988 Prediction from temperature of eyeing, hatching and 'swim up' times for salmonid embryos. *Freshwater Biology* 19, 41-8.

DONALD, A.P. and STONER, J.H. 1989 The quality of atmospheric deposition in Wales. *Archives of Environmental Contamination and Toxicology* 18, 109-19.

EGGLISHAW, H.J. 1986 Afforestation and fisheries. In J. de le Solbe (ed.), *Effects of land use on fresh waters*, 236-44. Chichester. Ellis Horwood.

ELLIOTT, J.M. 1975 The growth rate of brown trout (*Salmo trutta*) fed on maximum rations. *Journal of Animal Ecology* 11, 647-72.

ELLIOTT, J.M., HUMPESCH, U.H. and MACAN, T.T. 1988 *Larvae of the British Emphemeroptera. A key with ecological notes.* FBA Scientific Publication Number 49. Ambleside. Freshwater Biological Association.

FRANCIS, I.S. and TAYLOR, J.A. 1989 The effect of forestry drainage operations on upland sediment yields: a study of two peat covered catchments. *Earth Surface Processes and Landforms* 14, 73-83.

GEE, A.S. and STONER, J.H. 1988 The effects of afforestation and acid deposition on the water quality and ecology of upland Wales. In M.B. Usher and D.B.A. Thompson (eds), *Ecological change in the uplands*, 273-87. Oxford. Blackwell.

HARRIMAN, R. and MORRISON, B.R.S. 1982 Ecology of streams draining forested and non-forested catchments in an area of central Scotland subject to acid precipitation. *Hydrobiologia* 88, 251-63.

HARRIMAN, R., MORRISON, B.R.S., CAINES, L.A., COLLEN, P. and WATT, A.W. 1987 Long-term changes in fish populations of acid streams and locks in Galloway, south west Scotland. *Water Air and Soil Pollution* 32, 89-112.

HORNUNG, M., REYNOLDS, B., STEVENS, P.A. and NEAL, C. 1987 Stream acidification resulting from afforestation in the U.K.: evaluation of causes and possible ameliorative measures. *Forest Hydrology and Watershed Management*, IAHS-AISH publication No. 167.

HURLBERT, S.T. 1984 Pseudo-replication and the design of ecological experiments. *Ecological Monographs* 45, 187-211.

HYNES, H.B.N. 1975 The stream and its valley. *Verhandlung der International Vereinigung für Limnologie* 19, 1-15.

JONES, D.H. 1986 The effect of afforestation on fresh waters in Tayside, Scotland. Zooplankton and other microfauna. *Hydrobiologia* 133, 223-35.

LIKENS, G.E. 1984 Beyond the shoreline: a watershed-ecosystem approach. *Verhandlung der International Vereinigung für Limnologie* 22, 1-22.

LIKENS, G.E., BORMAN, F.H., PIERCE, R.S., EATON, J.S. and JOHNSON, N.M. 1977 *Biogeochemistry of a forested ecosystem*. New York. Springer-Verlag.

METCALFE, S.E. and DERWENT, R.F. 1990 Llyn Brianne: modelling the influence of emission reduction on deposition. In J.H. Stoner, A.S. Gee and R.W. Edwards (eds), *Acid Waters in Wales*, 299-310. The Hague. Kluwer.

MILLER, H.G. 1984 Deposition-plant-soil interactions. *Philosophical Transactions of the Royal Society of London* B305, 339-52.

NILSSON, S.I., MILLER, J.D. and MILLER, J.D. 1982 Forest growth as a possible cause of soil and surface water acidification: an examination of the concepts. *Oikos* 39, 40-9.

ORMEROD, S.J., and EDWARDS, R.W. 1985 Stream acidity in some areas of Wales in relation to historical trends in afforestation and the usage of agricultural limestone. *Journal of Environmental Management* 20, 189-97.

ORMEROD, S.J. and TYLER, S.J. 1987 Dippers *Cinclus cinclus* and Grey Wagtails *Motacilla cinerea* as indicators of stream acidity in upland Wales. In A.W. Diamond and F.L. Filion (eds), *The Value of Birds*, 191-208. Cambridge. International Council for Bird Preservation.

ORMEROD, S.J. and WADE, K.R. 1990 The role of acidity in the ecology of Welsh lakes and streams. In R.W. Edwards, J.H. Stoner and A.S. Gee (eds), *Acid Waters in Wales*, 93-119. The Hague. Kluwer.

ORMEROD, S.J., ALLINSON, N., HUDSON, D. and TYLER, S.J. 1986 The distribution of breeding dippers (*Cinclus cinclus*, Aves) in relation to stream acidity in upland Wales. *Freshwater Biology* 16, 501-8.

ORMEROD, S.J., BOOLE, P., McCAHON, C.P., WEATHERLEY, N.S., PASCOE, D. and EDWARDS, R.W. 1987 Short-term experimental acidification of a Welsh stream: comparing the biological effects of hydrogen ions and aluminium. *Freshwater Biology* 17, 341-56.

ORMEROD, S.J., BULL, K., CUMMINS, C., TYLER, S.J. and VICKERY, J.A. 1988 Egg mass and shell thickness in dippers *Cinclus cinclus* in relation to stream acidity in Wales and Scotland. *Environmental Pollution* 55, 107-21.

ORMEROD, S.J., DONALD, F.P. and BROWN, S.J. 1989 The influence of plantation forestry on the pH and aluminium concentration of upland Welsh streams: a re-examination. *Environmental Pollution* 62, 47-62.

ORMEROD, S.J., MAWLE, G.W. and EDWARDS, R.W. 1987 The influence of forest on aquatic fauna. In J.E. Good (ed.), *Environmental aspects of forestry production in Wales*, 37-49. ITE symposium number 22. Grange over Sands. Institute of Terrestrial Ecology.

ORMEROD, S.J., WADE, K.R. and GEE, A.S. 1987 Macro-floral assemblages in upland Welsh streams in relation to acidity and their importance to invertebrates. *Freshwater Biology* 18, 545-58.

ORMEROD, S.J., WEATHERLEY, N.S. and MERRET, W.J. 1990 The influence of conifer plantations on the distribution of the Golden Ringed Dragonfly *Cordulegaster boltonii* in the upper catchment of the River Tywi. *Biological Conservation* 53, 241-51.

ORMEROD, S.J., WEATHERLEY, N.S., VARALLO, P.V. and WHITEHEAD, P.G. 1988 Preliminary empirical models of the historical and future impact of acidification on the ecology of Welsh streams. *Freshwater Biology* 18, 127-40.

PARRY, M. and SINCLAIR, G. 1985 *Mid Wales uplands study*. Cheltenham. Countryside Commission.

REUSS, J.O., COSBY, B.J. and WRIGHT, R.F. 1987 Chemical processes governing soil and water acidification. *Nature* 329, 27-32.

REYNOLDS, B., NEAL, C., HORNUNG, M. and STEVENS, P.A. 1986 Baseflow buffering of stream acidity in five mid-Wales catchments. *Journal of Hydrology* 187, 167-85.

REYNOLDS, B., NEAL, C., HORNUNG, M., HUGHES, S. and STEVENS, P.A. 1988 Impact of afforestation on the soil solution chemistry of stagnopodzols in mid-Wales. *Water Air and Soil Pollution* 38, 55-70.

RUTT, G.P., WEATHERLEY, N.S., and ORMEROD, S.J. 1989 Microhabitat availability in Welsh moorland and forest streams as a determinant of macroinvertebrate distribution. *Freshwater Biology* 22, 247-61.

STONER, J.H. and GEE, A.S. 1985 Effects of forestry on water quality and fish in Welsh lakes and rivers. *Journal of the Institute of Water Engineers and Scientists* 39, 27-45.

STONER, J.H., WADE, K.R. and GEE, A.S. 1984 The effects of acidification on the ecology of streams in the upper Tywi catchment in west Wales. *Environmental Pollution (Series A)* 35, 125-57.

SULLIVAN, T.J., DRISCOLL, C.T., EILERS, J.N. and LANDERS, D.H. 1988 Evaluation of the role of sea salt inputs in the long-term acidification of coastal New England lakes. *Environmental Science and Technology*, 22, 185-9.

UNDERWOOD, J., DONALD, A.P. and STONER, J.H. 1987 Investigations into the use of limestone to combat acidification in two lakes in west Wales. *Journal of Environmental Management* 24, 29-40.

VICKERY, J.A. 1989 The effects of surface water acidification on riparian birds with particular reference to the Dipper. Unpublished D.Phil. thesis, University of Oxford.

WADE, K.R., ORMEROD, S.J. and GEE, A.S. 1989 Classification and ordination of macroinvertebrate assemblages to predict stream acidity in upland Wales. *Hydrobiologie* 171, 59-78.

WARREN, S.C., ALEXANDER, G.C., BACHE, B.W. *et al.* 1988 *United Kingdom Acid Waters Review Group Final Report*. London. Department of the Environment.

WARREN, S.C., BACHE, B.W., EDMUNDS, W.M. *et al.* 1986 *Acidity in United Kingdom Fresh Waters. United Kingdom Acid Waters Review Group Interim Report*. London. Department of the Environment.

WEATHERLEY, N.S. 1988 Liming to mitigate acidification in freshwater ecosystems: a review of the biological consequences. *Water, Air and Soil Pollution* 39, 421-37.

WEATHERLEY, N.S. and ORMEROD, S.J. 1990 Forests and the temperature of upland streams: a modelling exploration of the biological effects. *Freshwater Biology* 24, 109-22.

WEATHERLEY, N.S., RUTT, G.P. and ORMEROD, S.J. 1989 Densities of benthic macroinvertebrates in upland Welsh streams of different acidity and land use. *Archiv für Hydrobiologie.* 115, 417-31.

WELLS, D.E., GEE, A.S. and BATTARBEE, R.W. 1986 Sensitive surface waters: a UK perspective. *Water Air and Soil Pollution* 31, 631-68.

WELLS, D.E. and HARRIMAN, R. 1987 Acidification studies in Scottish catchments: the effect of deposition, catchment type and runoff. In R. Perry, R.M. Harrison, J.N. Bell and J.N. Lester (eds), *Acid Rain: Scientific and Technical Advances*, 293-300. London. Selper Limited.

Part 5

River management

In: Steer, M.W. (ed.) 1991 *Irish Rivers : Biology and Management*, pp 187-204. Royal Irish Academy, Dublin.

REHABILITATION OF SALMONID HABITATS IN A DRAINED IRISH RIVER SYSTEM

Martin O'Grady

Central Fisheries Board, Glasnevin, Dublin

INTRODUCTION

A post-drainage fishing rehabilitation programme for the Boyne catchment has been funded in recent years by the Office of Public Works. This paper describes rehabilitation procedures undertaken within this catchment.

The first step in this process was to undertake a detailed biological survey of the drained section of the catchment to establish the *status quo* post-drainage and thereafter, and to design rehabilitation programmes to restore the natural balance in the system.

This is an ongoing exercise. To date (February 1989), only 60% of the drained catchment area has been surveyed. Rehabilitation plans have been drawn up for many sections of this area. It is intended that the survey programme be completed in 1989 and rehabilitation works finalised by 1991.

THE BOYNE CATCHMENT

1. General background information

The Boyne is one of Ireland's larger catchments. It drains an area of c. 2,500 km (965 square miles). There are eleven major tributary catchments (Fig. 1).

Geologically, limestone formations dominate almost the entire catchment - even the extensive bog areas in the upper area of the catchment are underlain with limestone glacial drift. Riverine systems draining limestone formations are usually productive and the Boyne is no exception.

The gentle slopes throughout the Boyne catchment result in an almost continuous riffle/glide/pool sequence in both its tributaries and main channel. This natural pattern is interrupted in places by large weirs, particularly on the main channel. This type of system is ideally suited to supporting mixed stocks of salmon and trout.

While these two species tend to dominate populations in most sections of

188

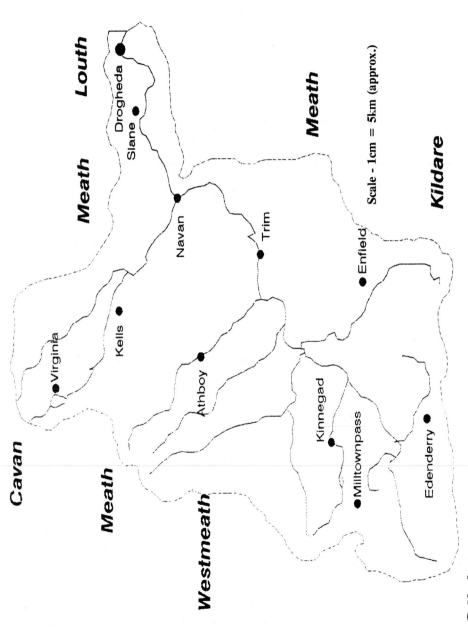

Fig. 1. The Boyne catchment.

the catchment they are not the only ones present - a total of thirteen species have been identified to date - salmon, trout, stoneloach, gudgeon, minnow, rudd, roach,bream, tench, 3-spined sticklebacks, freshwater eel, stream lamprey and flounder.

2. The distribution within and utilisation of this catchment by salmonid stocks

The salmonid populations use this catchment as a single integrated system. Presently the areas surveyed function, in this regard, in the following way.

i   - Spawning sites for salmon and trout are confined almost exclusively to the tributary catchments - there are very limited gravel deposits in the main channel.

ii   - The tributaries function as nursery areas for the two salmonid species and the larger of these channels also support substantial populations of resident brown trout.

iii   - Many main channel sections also support large adult brown trout stocks and significant salmon parr populations, the vast majority of whom are born in the tributary systems.

3. The extent of the angling resource

When the rehabilitation programme is complete (1992/1993) one can expect that there will be c. 213 km (123.2 miles) of trout angling channel in the catchment.

The salmon angling water is currently confined almost exclusively to channel areas downstream of Navan. This is because very few adult salmon move upstream of this point before the end of the angling season. There is no way of predicting whether or not this will be a long-term trend. There is some evidence to suggest that as long as salmon runs are dominated by the present large summer and autumn runs angling will be confined to the Navan and Drogheda zone except in years when there are exceptionally large summer floods.

This confinement of salmon angling to the lower reaches of the main channel is not unusual in an Irish context in recent years. A very similar situation has been observed in the Suir catchment since 1980 with the prime angling zones for this species being confined principally to channel areas downstream of Newcastle (T. Sullivan, pers. comm.).

DETAIL IN RELATION TO THE ARTERIAL DRAINAGE PROGRAMME

This programme commenced in 1969 and continued until 1985. Thereafter the only dredging activities ongoing were and are a part of the

190

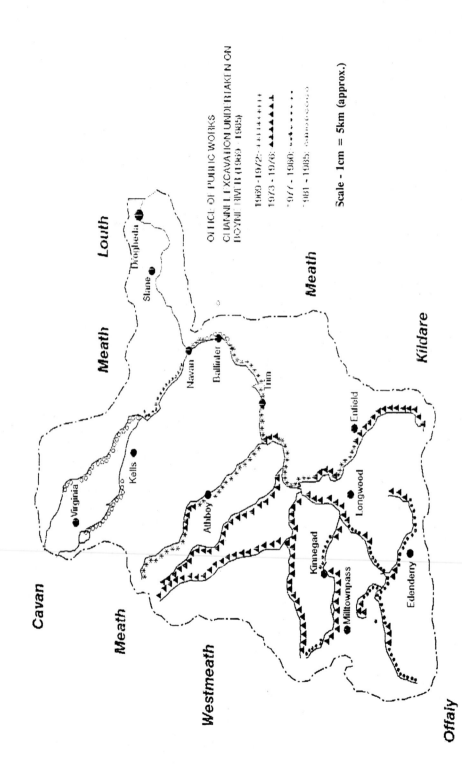

Fig. 2. Boyne catchment drainage scheme.

maintenance programme or the fishery rehabilitation exercise. Most of the former operations are undertaken in land drains which are of little or no value in fishery terms.

The rate of drainage works over the seventeen-year period are illustrated for each four-year period within the scheme (five years in relation to the end of the scheme) (Fig. 2). The side channels of the tributary catchments are not included because of illustration difficulties. However, practically all of these were also drained. No drainage works were undertaken in channels downstream of Navan and little work was carried out in the Kells Blackwater system upstream of Lough Ramor (Fig. 2). Generally speaking, no more than one sub-catchment was drained in any one year during the course of the scheme.

AN OVERVIEW OF THE EFFECTS OF THE DRAINAGE PROGRAMME ON SALMONID STOCKS IN THE CATCHMENT

In an attempt to obtain an overview of the long-term effects of drainage on the system, the commercial salmon catches in the Boyne were compared with those of the River Slaney, a channel which has not been subject to arterial drainage. Over the last hundred years both of these catchments have been recognised as important east coast salmon fisheries. In making this comparison the author has assumed that both systems were subject to similar levels of agricultural pollution and that salmon produced in both catchments were exploited to an equal extent by the high seas drift-net fishery.

The figures used were the official returns published in the annual reports of the Department of the Marine. The values compared graphically were the C.P.U.E. (catch per unit effort) figures calculated per annum, i.e. the total numerical annual commercial salmon catch in traps and drift nets divided by the number of licensed fishermen (traps and drift nets) in that year (Fig. 3). Commercial salmon catches in the Department of the Marine reports are presented in terms of total weight. The number of fish involved was estimated by dividing this total weight value by the mean (X) weight of rod-caught fish in each river in the relevant year. The latter values were calculated from the total number and weight values presented in the D.O.M. reports for rod-caught fish in the relevant catchments.

These data indicate a relatively close and parallel pattern in commercial salmon catches for both systems since records began (1947 up to 1987). In 1988 there was a very significant increase in the Boyne C.P.U.E. value relative to the Slaney (Fig. 3).

For the period 1947 to 1988 C.P.U.E. values were higher in the Slaney than the Boyne in ten individual years, five pre-drainage and five post-drainage. A statistical comparison of these C.P.U.E. values for both rivers has been carried out by Mr T. Joyce, B.E., Office of Public Works, Hydrological Section. He found (pers. comm.) that when comparing mean (X) C.P.U.E. ratio values for the period 1970 to 1988 the Boyne has, in terms of salmon returns, outperformed the Slaney by 17.6% in the years post-drainage (Table 1).

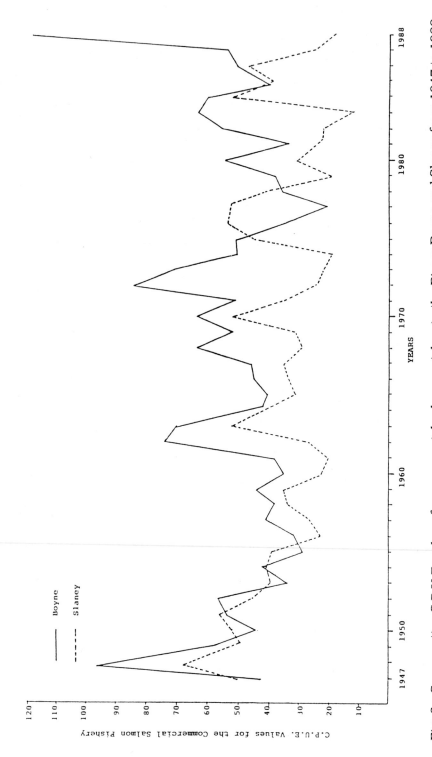

Fig. 3. Comparative C.P.U.E. values for commercial salmon catches in the Rivers Boyne and Slaney from 1947 to 1988 inclusive. C.P.U.E. = Mean ($\bar{x}$) number of salmon caught per licence per annum.

Table 1. Performance of short-term mean C.P.U.E. ratios.

Mean CPUE 1947-'69	50.90	38.11	1.336	1 : 1
DATE	BOYNE	SLANEY	RATIO	PERFORMANCE
1970	64.75	51.90	1.248	0.934
1971	57.91	43.13	1.343	1.005
1972	66.07	37.49	1.762	1.319
1973	67.28	33.95	1.982	1.484
1974	63.87	31.18	2.048	1.553
1975	61.64	33.54	1.838	1.376
1976	57.74	36.62	1.577	1.180
1977	53.22	38.77	1.373	1.028
1978	51.30	39.11	1.312	0.982
1979	49.91	36.99	1.350	1.010
1989	50.24	36.43	1.380	1.330
1981	48.88	35.27	1.386	1.374
1982	49.39	34.27	1.441	1.079
1983	49.90	32.70	1.526	1.142
1984	50.22	34.15	1.471	1.101
1985	49.34	34.39	1.435	1.074
1986	48.88	35.14	1.391	1.041
1987	49.29	34.57	1.426	1.067
1988	52.87	33.65	1.571	1.176

From Mr T. Joyce, Office of Public Works, Internal Report.

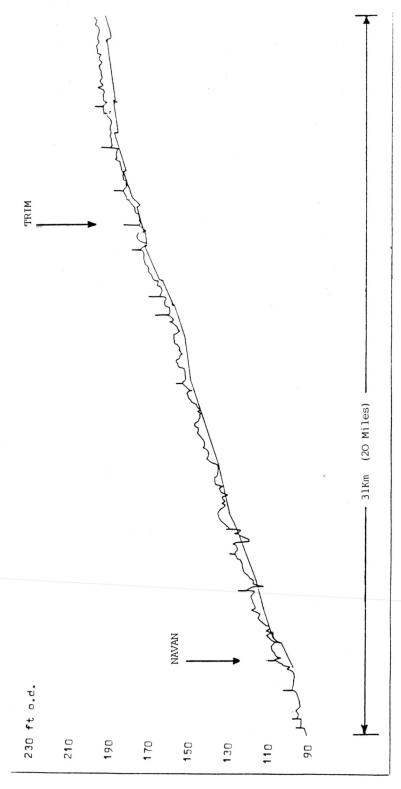

Fig. 4. Long section of the Boyne channel from Trim to Navan pre- and post-drainage (supplied by J. Curtain, B.E., Office of Public Works). The vertical strokes illustrate the position and relative height of the weirs.

It is widely acknowledged that the high seas drift-net fishery has been taking a greater proportion of returning adult salmon in recent years (Anon. 1986; 1987). It is not therefore surprising that returns to the estuarine salmon fishery in the Slaney have declined in recent years as illustrated by the C.P.U.E. values presented. The fact that this trend is not evident in the Boyne suggests that the drainage programme did not reduce salmon production in the system. In fact data suggest that drainage works may have actually resulted in increased salmon production in this river. The available evidence in relation to cropping rates of Boyne salmon by the high seas drift-net fishery suggests that fish from this catchment are probably exploited to at least the same extent as those from other Irish east coast riverine systems (J. Browne, pers. comm., Fisheries Research Centre, Abbotstown). Consequently the Boyne C.P.U.E. values in recent years may indicate an increased production of salmon in this catchment in the post-drainage era.

Factors responsible for the maintenance of good salmon stocks in the Boyne post-drainage

A number of factors are probably responsible for maintaining good salmon stocks in the system post-drainage.

i. A limited programme

The lower reaches (30 km) of the main Boyne channel (Navan to Drogheda) were never drained. Fish stocks in this area have never been quantified. However, there are significant lengths of fast-flowing shallow glides in this zone. Such areas, upstream of Navan, presently support large salmon parr populations. Consequently there was probably a continuous production of salmon downstream of Navan during drainage works.

ii. Change in bed levels

During the course of drainage works a total of eleven substantial weirs were removed from the channel from a point upstream of Trim downstream to Navan, a distance of 31 km (Fig. 4). The engineering longitudinal section for this area indicates that the weirs in situ pre-drainage "ponded" long sections of channel. Channel lengths not influenced by the weirs were for the most part very deep and slow-flowing (Fig. 4).

The ponding effect of the weirs allowed large quantities of silt to build up along the banks and in mid-stream positions to form islands.

Data available from a number of sources in relation to fish stocks in this zone pre-drainage confirm that it did not accommodate a significant population of salmon parr. A. McGurdy (currently Manager of the Eastern Regional Fisheries Board) was a crew member on electrofishing operations in this area in the 1950s designed to cull pike. He recalls the presence of a small population of adult trout, some pike and adult salmon in this area at that time (pers. comm.). Anglers living in the Trim area (J. Goggins, pers. comm.) confirm Mr Gurdy's observations. In addition Mr John Browne, Fisheries

Biologist, Dept. of the Marine, who was familiar with sections of this channel pre-drainage confirms that it was ecologically unsuitable as salmon nursery water (pers. comm.).

Electrofishing data compiled for this channel section in 1986 and 1988 suggest that it is now capable of producing c. 30,000 smolts annually.

McCarthy (1977; 1983) and J. Browne (pers. comm.) quantified salmonid stock levels for sites in a range of Boyne tributaries pre-drainage and for a period of two to five years after drainage. Their data indicate a significant reduction in salmonid stock densities in the immediate (2-5 year) post-drainage period at some sites. It would appear, therefore, that the creation of productive salmon nursery water in the main Boyne channel area from Trim to Navan may have prevented the collapse of the salmon stock in the short term and, as individual tributaries recovered from the effects of drainage, this additional main channel nursery zone may have increased the salmon production figure for this river system as a whole to a higher level than was present pre-drainage.

iii. Desilting the main channel during drainage works

When sections of the main Boyne channel were being drained considerable amounts of heavy silt were disturbed and washed downstream for a short distance before settling out again. These silt beds had accumulated in the channel over 100 to 200 years by the ponding effect of the large weirs. An overlapping sequence in dredging operations was a design feature of the scheme. This ensured that silt beds created in the channel by dredging works in year one were removed in year two by a machine which commenced operations downstream of where dredging started in year one. In this way large silt deposits which would have created an unproductive medium in salmonid terms were removed from the channel. Data compiled on the nature of the river bed in recent years (1986-1988) confirm the absence of significant silt deposits over the entire channel length from Trim to Navan (31 km) (O'Grady, unpubl.).

iv. The duration of the drainage scheme

The entire scheme was carried out over seventeen years. Details of the work schedule have already been presented (page 191). The longevity of the scheme must also have helped to prevent a collapse of salmon stocks in the short term with the undrained sections continuing to produce fish in the early years of works and channels drained early in the scheme recovering as other areas were drained.

THE POST-DRAINAGE REHABILITATION PROGRAMME

To date c. 60% of all drained channel areas have been surveyed. It is hoped that all remaining drained areas can be examined in 1989.

Procedure adopted has been as follows.

i   - All channels >3m in width have been electrofished qualitatively throughout their length. Smaller channels were electrofished at intervals.

ii   - The following information was also compiled for all channels surveyed: channel depth, width, substrate type, % pool/glide/riffle in all zones, aquatic floral and faunal data, and water analysis. In addition, the height and nature of banks and their flora were recorded. Engineering records for the long section of channels pre - and post-drainage were also examined. All of these data were used to construct an ecological profile of individual channels.

iii   - Quantitative electrofishing data were compiled in the more extensive ecological zones as defined using the terms of reference in (ii).

iv   - A rehabilitation programme was drawn up.

1. Problems identified and the solutions adopted to resolve them

Four major problem areas have been identified to date. Pilot rehabilitation works were undertaken in these specific areas to ensure that any rehabilitation programmes undertaken on a large scale were likely to prove successful.

1.1 The riffle/glide/pool sequence

Post-drainage the natural riffle/glide/pool sequence in sections of some tributary channels had been physically disturbed, being replaced by a continuous shallow glide sequence. Electrofishing data indicated that such zones tended to support small to moderate numbers of salmon parr and small (< 25cm) trout. A number of such areas were altered by the construction of weirs, pair and single arm deflectors, and adjacent control areas were left unaltered. A monitoring of stocks in such areas over the last four years indicates a consistent increase in salmonid numbers in the altered channel areas compared to the control zones - a two- to fourfold increase on average in both salmon parr and trout numbers was achieved. These increases represent a return of salmonid stock density figures to within the range one would expect in undrained Irish limestone streams of this type.

The success of this type of operation is not surprising. It has been used by many workers in the field throughout the world.

The speed at which stocks increased post-works surprised the author - optimum stocks appeared to have become established only one year after works and stabilised thereafter. This may be due to the fact that the river in question had been drained sixteen years prior to rehabilitation works and had an established and diverse macrophyte flora and invertebrate fauna

Table 2. A comparison of juvenile salmonid numbers in two discrete and adjacent areas of the Curraghtown Stream (Moynalty catchment).

Site   A uniform sandy bed with trailing grasses along both banks and no treeline. No in-stream flora.

Description	Mean (x) stream width	-	3.5 m
	Mean (x) depth	-	0.35 m
	Area fished	-	350 m^2
	Range of stream flows	-	20 to 35 cm/sec (low summer level)

	Salmon	Trout
Salmonid stocks nos/m^2	0.001	0.03

[for 0+ and 1+ year-old fish]

———————

Site   A rubble-strewn bed colonised with mosses. Stones ranged in size from 10 cm to 30 cm in diameter. Trailing grasses along both banks and no treeline.

Description	Mean (x) stream width	-	3.5 m
	Mean (x) depth	-	0.35 m
	Area fished	-	345 m^2
	Range of stream flows	-	2 to 51 cm/sec (low summer level)

	Salmon	Trout
Salmonid stocks nos/m^2	0.53	0.61

[for 0+ and 1+ year-old fish]

prior to physical alterations taking place. These data indicate that the presence of a complex hydraulic regime in terms of variable flows and channel depths plays a critical role in determining the salmonid carrying capacity of channels. This particular study will be the subject of a subsequent publication (O'Grady and Caffrey in prep.).

1.2 Compacted smooth bed areas

It was noted during the survey stage that smooth bed areas lacking floral colonies in the smaller (<5 m) streams held significantly less fish than rubble-strewn zones. These unproductive bed types were comprised of various materials - compacted clays, smooth rock, sand or marl. The level of differences in salmonid stock levels in smooth and rubbled areas are illustrated in Table 2 for two adjacent sections of a small stream in the Moynalty catchment.

A "rubbling" programme is being undertaken in such areas as part of the rehabilitation programme wherever possible.

1.3 Channel constrictions

The overlapping dredging sequence already described in relation to drainage works on the main Boyne channel did not take place to any extent in tributary catchments. Here, a machine commenced at the outfall of the channel to the Boyne and worked upstream towards the source. In a few tributaries this procedure allowed disturbed silt to wash downstream and lodge in areas of poor gradient. In places these deposits settled out along the banks, thereby creating an artificially narrow deep channel. Large deposits of silt in a few places tended to choke almost the entire channel width. All of these disturbed silt deposits were quickly colonised with emergent aquatic and semi-aquatic vegetation species which bound them and prevented a flushing of the silt beds in subsequent years.

The areas directly affected by silt deposits were found to be virtually barren in terms of salmonid populations. In addition, the larger silt deposits functioned as weirs, ponding channel areas upstream of the silt deposits up to 2 km in length. These ponded zones also held negligible salmonid stocks.

In the 1988/89 period a number of these silt deposits will be removed. Initial observations suggest that this desilting exercise will prove very productive. Channel bed areas in most zones following desilting have a gravel/sandy/stony bed which should provide the basis for a productive salmonid habitat.

The length of gravelled channel in the entire Boyne tributary catchment was estimated for both the pre- and the post-drainage era (R. Flusky, F.R.C., pers. comm.). Data indicate a significant increase in the extent of exposed gravel beds post-drainage - an overall 15% increase in gravelled channel length in the tributary catchments.

To date, one such gravel bed in the Deel River was found to be detrimental to fishing interests. This excessive deposit, washed into the channel from eroding banks post-drainage, was found to be ponding a significant length of

Table 3. Initial results for the colonisation by salmonids of pilot rehabilitation areas in the main Boyne channel.

	Area fished (m²)	Prior to development		After development		Time-span after development
		Salmon parr	Trout	Salmon parr	Trout	
Stone bands						
Band series 1 (Scariff site)	2,616	0 (0)	2 (5x10⁻⁴)	98+/-65 (0.02)	25+/-6 (0.006)	5 months
Band series 2 (Inchamore site)	3,888	0 (0)	1 (2.6x10⁻⁴)	80+/-21 (0.02)	42+/-5 (0.01)	12 months
Control area	3,888	0 (0)	1 (2.6x10⁻⁴)	0 (0)	0 (0)	12 months
V rubble structures						
Expt. area	3,200	101+/-33 (0.03)	20+/-3 (0.006)	107+/-9.5 (0.03)	79+/-8 (0.02)	5 months
Control area	3,200	140+/-26 (0.04)	25+/-7 (0.08)	161+/-23 (0.05)	18+/-3 (0.05)	5 months

( ) - Mean (x) number of fish per square metre.

channel (*c.* 0.5 km).

1.4 Problems in the main Boyne channel

In 1986, 36.4 km of the main Boyne channel (drained section) were electrofished - from Longwood downstream to Ballinter House. Data compiled
indicate clearly that substantial salmonid stocks were present in all relatively
shallow (<0.7 m) fast-flowing zones where there was a complex flow regime
varying from back eddies to fast discharges (160 cm/sec) (low summer flow
values).

In slower-flowing (20-25 cm/sec) moderately deeper (<1.5 m) areas where
the discharge regime was relatively uniform across the channel width, very
poor stocks of trout and salmon parr were present despite the fact that a relatively complex and abundant flora and invertebrate fauna were present in
such areas. In deeper (1.5-2.5 m) slower-flowing (10-15 cm/sec) areas, virtually no salmonids, except occasional adult salmon, were present. Salmonid
stocks were not replaced by cyprinid populations in such areas probably
because the river bed is of a sandy/stony/gravelly nature and therefore incapable of supporting substantial chironomid or tubificid populations, the
major dietary items of cyprinid fishes.

Pilot rehabilitation works were undertaken in both the moderately deep
and very deep zones described above in 1987 and 1988. Essentially they
involved the placement of stone deposits on the river bed to create hydraulic
conditions similar to those observed in the shallow areas which were found
to support substantial salmonid stocks. Two types of "structure" were introduced. They are illustrated schematically in Fig. 5. Two band type series
were constructed in very deep (2.0 m) channel areas and the alternate regime
was introduced at a third site. These three areas and two appropriate control zones were electrofished both prior to works and again 5 months (2 sites)
or 12 months (1 site) after works were complete. These provisional data compiled post-works suggest that this type of programme will prove successful in
generating salmonid stocks in the long term (Table 3). Stocks of salmon
parr and trout increased dramatically in the areas where bands were constructed. At the site where V-shaped rubble structures were introduced a
moderate salmon parr stock present pre-works was maintained and a very
significant increase in trout stocks was achieved. The increased trout stocks
at all sites encompassed a range of individual fish from 10 cm to 45 cm in
length.

The major increases in salmonid stocks in these areas after such a short
period in zones where a permanent floral and invertebrate faunal regime has
not yet become established suggests that further increases in density of
stock levels can be expected in the long term. It is also possible that the
mid-stream deflector/rubbled margin system may support salmonid fry in
time and perhaps even accommodate spawning salmon because the marginal
rubbled areas contain significant quantities of larger gravels. These data
also indicate both the importance of a complex flow regime in providing salmonid "lies" and the ability of these fish to survive on invertebrate drift in the
summer/autumn period.

202

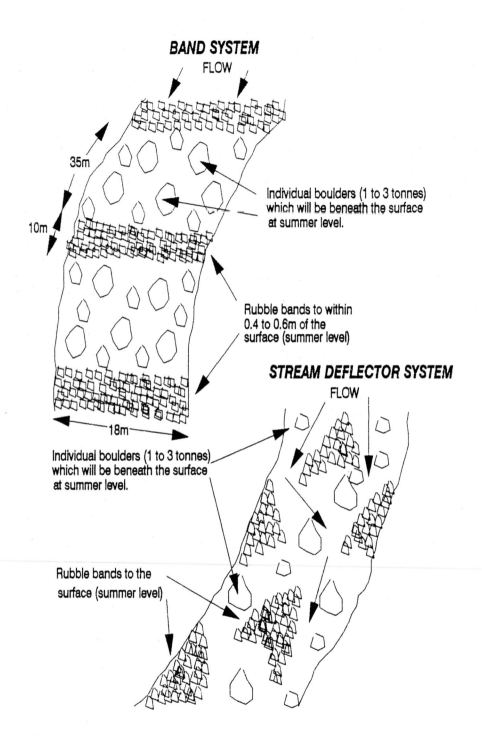

BAND SYSTEM

FLOW

35m

10m

Individual boulders (1 to 3 tonnes) which will be beneath the surface at summer level.

Rubble bands to within 0.4 to 0.6m of the surface (summer level)

STREAM DEFLECTOR SYSTEM

FLOW

18m

Individual boulders (1 to 3 tonnes) which will be beneath the surface at summer level.

Rubble bands to the surface (summer level)

Fig. 5. Schematic representations of the alternate rubble systems.

It is intended to monitor these sites over the next three to four years to establish the long-term effects of these structures on the ecology of the channel and in particular on the level of salmonid stocks present.

CONCLUSIONS

Available data on salmon catch statistics suggest that the Boyne may be producing as many, if not more, salmon post-drainage than it did in the pre-drainage era. The removal of eleven large weirs, and the silt accumulated by these structures in the main Boyne over the last two centuries, has created many extensive shallow, moderately fast-flowing areas suited to trout and salmon parr over a 31 km length of channel from Trim to Navan.

Survey data indicate that many sections of tributary catchments have recovered naturally post-drainage in terms of being able to support substantial salmonid stocks.

A monitoring of the rehabilitation works carried out to date suggest that they will significantly enhance salmonid stocks.

It is likely, therefore, that with the budget available for rehabilitation works the Boyne will, in the long term, be a significantly more productive salmonid system than it has been in the immediate past.

ACKNOWLEDGEMENTS

Special thanks are due to the Office of Public Works (Arterial Drainage Section) who funded this study. In particular the author would like to thank John Howard, B.E., Assistant Chief Engineer, who initiated this programme. I am most grateful to John Curtain, B.E. (O.P.W., Trim), whose engineering experience was critical in terms of designing rehabilitation programmes capable of enhancing salmonid stocks without interfering with the drainage function of this channel. Thanks are also due to Tim Joyce, B.E. (O.P.W.), for his statistical analysis of commercial salmon catch figures.

I would like to thank Inspector John Stapleton and his staff of the Eastern Regional Fisheries Board for all of their assistance.

Special thanks are due to Richard Flusky (F.R.C.), John Stapleton (E.R.F.B.) and Alan Williams, B.E. (Department of the Marine), for their assistance in drawing up the various rehabilitation schemes and to Mr Alan McGurdy (E.R.F.B.), John Brown (F.R.C.) and Tom Sullivan (S.R.F.B.) for their personal communications.

I am grateful to Paddy Fitzmaurice, Ph.D. (C.F.B.), for his helpful comments in relation to the preparation of the manuscript.

REFERENCES

ANON. 1986 *Inland Fisheries. Strategies for Management and Development.* Dublin. Central Fisheries Board.

ANON. 1987 *Report of the Salmon Review Group. Framework for the Development of Ireland's Salmon Fishery.* Dublin. Dept. of the Marine.

204

McCARTHY, D.T. 1977 The effects of drainage on the Trimblestown River. I Benthic invertebrates and flora. *Irish Fisheries Investigations* Ser. A., No. 16.

McCARTHY, D.T. 1983 The impact of arterial drainage on fish stocks in the Trimblestown River. *Irish Fisheries Investigations* Ser. A., No. 23.

O'GRADY, M.F. and CAFFREY, J. (in prep.) The effectiveness of a post drainage fishery rehabilitation programme in the Stonyford River.

In : Steer, M.W. (ed.) 1991 *Irish Rivers : Biology and Management*, pp 205-218.
Royal Irish Academy, Dublin.

ARTERIAL DRAINAGE, STREAMFLOW
AND THE RIVER ENVIRONMENT -
SOME EXAMPLES FROM NORTHERN IRELAND

David N. Wilcock

Department of Environmental Studies, University of Ulster

and

Charles I. Essery

Department of Computer Science, University of Ulster

ABSTRACT

Arterial drainage (or channelisation) is undertaken to facilitate
improved land drainage, to eliminate flooding, and to enhance agricultural output. Since 1945 over 5,000 km of "main" and "minor"
river channel in Northern Ireland have been channelised. The process involves widening, deepening and straightening rivers and
results in modifications to the river environment in a variety of ways.
Important instream habitats are lost, streamflow characteristics
transformed, suspended sediment loads increased and wetland areas
of the floodplain changed in character. An instrumented catchment
study on the River Main in County Antrim since 1978 allows many of
these environmental impacts to be quantified for the first time in
Northern Ireland.

INTRODUCTION

Arterial drainage or channelisation (Brookes 1988) is an integral element
of land drainage throughout Ireland and involves widening, deepening and
straightening river channels. In some major schemes old channels are completely filled in and new ones excavated in a completely new location away

P

from the original channel. The process ensures that the main arterial waterways of the country are able to accept and transport larger quantities of water than in their unchannelised condition without overflowing their banks. Channelisation thus helps eliminate flooding in low-lying floodplain areas where land is frequently of relatively high potential agricultural value. Deeper river channels may also provide a direct drainage benefit by lowering water-tables in immediately adjacent land, though the realisation of this benefit depends very much on soil permeability which throughout Ireland is frequently very low, even in floodplain areas. An indirect drainage benefit of deepened rivers is that they ensure unblocked outfalls from field drainage systems which, after channelisation, are able to operate more efficiently than before, particularly when the rivers are in spate.

In both the Republic of Ireland and Northern Ireland the most recent phase of drainage began in the late 1940s with the perceived need at that time for agricultural self-sufficiency. Given the water-retentive properties of many Irish soils, drainage was a prerequisite for any expansion or intensification of agriculture, and central government funding for land drainage has accordingly been generous. Since the early 1970s central government assistance has been supplemented by European funding of arterial drainage, most notably through the European Agricultural Guidance and Guarantee Fund. Given the current emphasis on taking land out of agriculture throughout Europe, the scale of arterial drainage may decline in the imminent future. Whether or not such a decline takes place the effects of channelisation on rivers throughout Ireland will remain for some time.

Between 1945 and 1984 ten major and four minor river systems in the Republic of Ireland have been drained (Kelly 1984). Bruton and Convery (1982) quantify the scale of land drainage over the same period. In Northern Ireland, over 5,000 km of main and minor river have been drained since the 1947 Drainage Act (Wilcock 1979). This is the equivalent of 0.34 km per square km, a figure higher than for any of the other countries quoted by Brookes (1985; 1988) except Denmark. The direct environmental effects of channelisation are mainly physical and affect channel geometry, streamflow, and suspended sediment loads. These impacts, however, in turn affect biological processes in and adjacent to the river. Loss of floodplain wetland, changes in the time distribution of low and high flows, changes in the magnitude of flood peaks and in their associated velocities, loss of instream habitat diversity, increases in suspended sediment concentrations; all these physical effects have "knock-on" biological consequences of importance for river management.

A preliminary quantification of some of these effects is now possible for the upper River Main in Northern Ireland. An arterial drainage scheme was started on this river in the middle 1970s following a Public Inquiry in 1971 (Hutton 1972). At the Public Inquiry conflicting evidence was presented as to the likely effects of the scheme on a whole range of river characteristics, notably fisheries, low flows, and flood flows. Since 1978 the upper 200 square kilometres of the River Main has been instrumented (Fig. 1) and records made of precipitation, streamflow, evapotranspiration, groundwater, and suspended sediment.

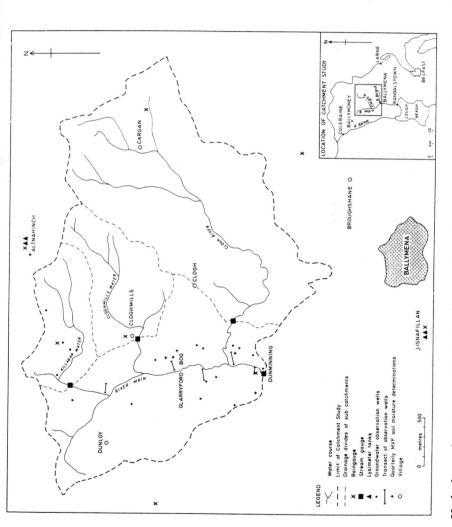

Fig. 1. Hydrological instrumentation for the study of channelisation in the upper River Main catchment, County Antrim, 1978-89.

Monitoring has been undertaken in four catchments defined by stream-flow gauges on the River Main at Dunminning and on three tributary catchments, the Killagan, the Cloghmills Water and River Clogh. Arterial drainage commenced at Ballymena, 15 km downstream of Dunminning, in the mid 1970s. Proceeding upstream, engineering work reached Dunminning in April 1984. Data for the five-year period before this date represent pre-channelisation conditions. Only three years' data for the period after 1984 have so far been analysed. This represents the post-channelisation period. Data from the three tributary catchments, all undrained, represent controls against which pre- and post-channelisation changes on the River Main can be assessed. By subtracting the sum of streamflows on the three left-bank tributaries from streamflow at Dunminning, it is possible to hydrologically isolate the area of the River Main catchment above Dunminning in which the impacts of channelisation are greatest. This area is called the Glarryford catchment.

Preliminary assessments of change in the river environment of the River Main as a result of channelisation are presented under the headings channel geometry, suspended sediment, and streamflow.

CHANNEL GEOMETRY

Changes to the planform in five of the sixteen km above Dunminning affected by channelisation are presented in Fig. 2. The original length of channel in this reach was 5.23 km. This was reduced by channelisation to 4.04 km, a reduction in channel length of 23%. The immediate effects of this shortening are twofold: a loss of meanders and of the pool and riffle habitats associated with them, and a steepening of channel gradient. At low flow, meander pools are low-energy environments and provide important food supplies and resting places for fish. At high flows they become high-energy environments at which surface streamflow converges before plunging to the base of the channel. The zone of *relatively* low energy is now transferred to the area of point bar sediment accumulation on the inside of the meander bend which may, therefore, become the preferred location for fish seeking refuge from high velocity streamflow during floods. Not only do these areas represent physical conditions of relative calm during flood flow but they may also receive food dredged up from the pools and transported to them by secondary flow circulation. The morphological and sedimentological diversity which characterises a meandering stream thus provides a range of habitats which enhances biological richness and diversity within relatively small areas. Realigned channels with straight banks and trapezoidal cross-sections contain no such diversity (Newbold *et al.* 1983; Lewis and Williams 1984; Brookes 1988).

Steepening of the channel slope is important because it tends to increase stream velocity. Several equations are used to relate velocity to channel morphology, all of them showing velocity to be directly related to slope and depth and inversely related to channel roughness or a friction factor.

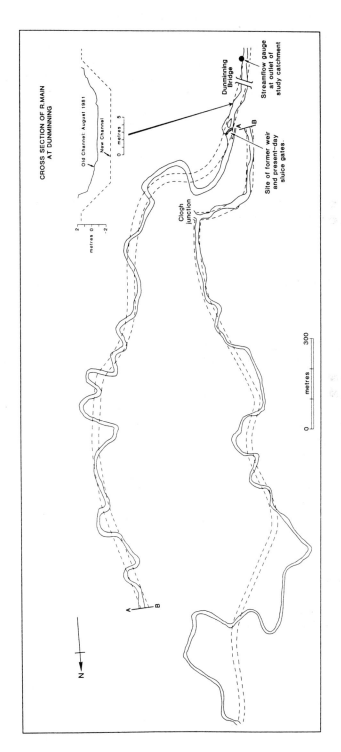

Fig. 2. Planforms of the River Main above Dunminning gauge before and after channelisation. Before and after cross-sections at Dunminning are also shown.

A widely-used equation governing velocity in streams is Manning's equation:

$$V = S^{0.5} \; R^{0.66}/n$$

where V is mean stream velocity, R is the hydraulic radius of the cross-section (i.e. cross-sectional area/wetted perimeter), S is slope, and n is a measure of channel roughness. In the length of river illustrated in Fig. 2 the new slope tangent (0.000218) is 1.3 times as steep as that for the pre-channelised slope (0.000168). This leads to a 14% increase of the term $S^{0.5}$ in the Manning equation.

Similar calculations can be made for the other two terms in Manning's equation. The channel cross-sections at Dunminning in pre- and post-channelisation conditions are also shown in Fig. 2. Bankful R following channelisation is 2.49 m, 69% greater than the pre-channelisation figure of 1.47 m. A change of this magnitude produces an increase of 43% in the R term as weighted in the Manning equation. This is probably not typical of increased R values along the length of river shown in Fig. 2, for overdeepening in the vicinity of Dunminning gauge has been greater than elsewhere in the channelisation scheme.

As regards increases in channel roughness, these have to be estimated. Wilson (1983, 301) quotes n values of 0.21 for channels in "rough-dressed stone paved, without sharp bends", and of 0.45 for rivers "with shallows and meanders and noticeable aquatic growth". Extrapolation of these figures to the River Main suggests an estimated reduction of roughness in the region of about 50% for the length of river shown in Fig. 2. The clear conclusion is that channelisation changes very considerably ALL factors which affect channel velocity. The changes, moreover, are all in a direction which leads to increased velocity. A potential ecological effect of increased velocities is on the movement of game fish out of and back into a river system. Downstream movements might be made easier, perhaps too easy for salmonids in the early part of their life cycle, while upstream movement of mature adults to spawning grounds might be inhibited. Any increases in velocity brought about by channelisation are only compounded, of course, if refuges for shelter are also reduced in number.

Biological studies that attempt to relate the occurrence of plant and animal species to velocity are rare, and those that are undertaken often refer only to the velocity obtained at the time of biological sampling. Alabaster (1970) attempted to relate upstream movement of salmon to river flow while accepting that streamflow might only be a surrogate for "prime stimuli" such as velocity and water quality. Plant and animal species, however, have to adapt to a range of velocities in any one habitat, infrequent high flows having flow velocities several times higher than more frequent low flows. Description of flow velocities in terms of their expected duration, therefore, more usefully describes habitat characteristics than isolated field measurements, and this has been attempted for pre- and post-channelisation conditions at the Dunminning site (Fig. 3). It is evident from Fig. 3 that mean instream velocities for all flow durations have increased, by a factor of between 2 and

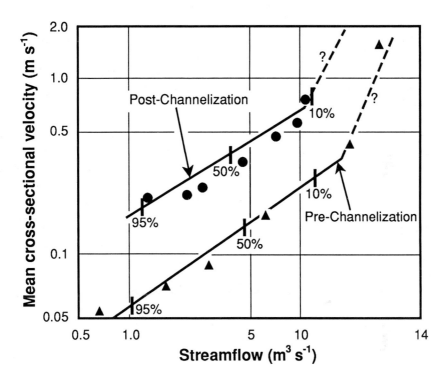

Fig. 3. Changes in mean velocity with streamflow for pre- and post-channelisation conditions; 5%, 50% and 95% duration flows are indicated.

Table 1. Mean cross-sectional velocities and flow duration in channelised rivers of Northern Ireland.

% time flow equalled or exceeded	Velocity (m/sec.)
95	0.03 - 0.30
50	0.31 - 0.40
10	0.42 - 1.0
5	0.63 - 1.58

4, as a result of channelisation, the largest relative increases being associated with the more extreme flows.

The data shown in Fig. 3 need to be qualified. Mean velocity is not stream-flow/cross-sectional area but rather the mean of measured velocities actually recorded in the field as part of the stream gauging operation. Mean velocity for the cross-sections as a whole would, therefore, be marginally lower than that shown and this would apply for both pre- and post-channelisation conditions. A second qualification is that data presented do not refer to flood conditions. A 10% flow is in fact a relatively small discharge and would occur as the crest of a flood peak during small spates only. Finally, it bears repetition that changes to channel geometry at Dunminning were particularly severe and relative increases in mean stream velocity shown in Fig. 3 as a result of channelisation most likely represent the maximum amount of change probable.

An examination of mean cross-sectional velocity in relation to flow duration has also been undertaken at 14 sites throughout Northern Ireland using data provided by the River Engineering Division of the Department of Agriculture for Northern Ireland. The sites examined relate to catchments ranging in drainage basin area from 64 square kilometres to 646 square kilometres, some of which have been channelised in the last 40 years. These data illustrate the velocity conditions that characterise lowland rivers in Northern Ireland at the present time. Though Table 1 excludes data from Dunminning, its general character is confirmed by field data collected there. The lower velocities in each of the ranges presented in Table 1 approximate to pre-channelisation conditions at Dunminning. The upper values appear more typical of post-channelisation conditions.

SUSPENDED SEDIMENT

At the Public Inquiry into the River Main Drainage Scheme (Hutton 1972), much of the debate focussed on the effects of channel excavation on the release of fine sediments, and the possible effects this would have on fisheries, the Main being one of the finest game fisheries in the Lough Neagh system. The deposition of fine sediments on top of coarser bed material has many documented effects. It may kill benthic invertebrates that require a coarse bedload environment. It may block the interstices between coarser bed material in which salmon lay their eggs, thus inhibiting the transfer through the bedload of essential oxygen-rich water and the removal of metabolic waste. Very high concentrations of suspended sediment may have an abrasive effect on fish (Swales 1982).

Given the very limited quantitative information on suspended sediment levels in non-channelised and channelised rivers in Ireland it was decided to monitor concentration at the Dunminning gauge. Two approaches were used. In the first phases of monitoring, depth-integrated suspended sediment samples were taken with each measurement of streamflow and rating curves of suspended sediment concentration (mg/litre) were plotted against measured streamflow (cubic metres /second). In the second, post-channelisation phase of study, continuous sampling was undertaken at two-hourly intervals.

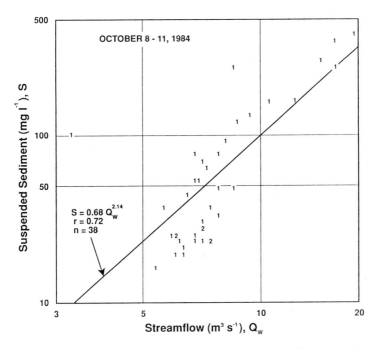

Fig. 4. Suspended sediment concentrations versus streamflow for an individual hydrograph, October 1984.

In pre-channelisation conditions 234 individual samples were collected and analysed. On each site visit three samples were taken at one-third, half, and two-thirds the distance across the river. Samples relating to a sample of 78 flow conditions are thus available for pre-channelisation conditions. The mean and median suspended sediment concentrations derived from these data are 15.8 mg/litre and 7.9 mg/litre respectively. The highest concentration recorded was 151.8 mg/litre, on the rising limb of a flood hydrograph. Only one other value exceeded 100 mg/litre in two years of recording between March 1981 and February 1983. These figures are very much in line with typical suspended sediment loads on British and Irish rivers (Gregory and Walling 1973; Lewin 1981; Carling 1988). During channelisation upstream of Dunminning suspended sediment levels increased considerably. From 52 miscellaneous readings taken between July 1984 and October 1985, i.e. in the first nineteen months of channelisation above Dunminning, the mean and median suspended sediment concentrations were 116.7 mg/litre and 54.6 mg/litre respectively. Fifteen samples had concentrations above 100 mg/litre, and four values were above 1000 mg/litre. The highest recorded value was 1210 mg/litre. These values are very similar to those described by McCarthy (1980).

From October 1984 an automatic peristaltic pump sampler was installed at Dunminning. Twenty-four samples a day can be determined from this device and rating curves of suspended sediment against streamflow can be established for individual flood hydrographs. Examination of these relationships for individual flood peaks in October 1984 and July 1985 illustrates the great variability in the relationships between streamflow and suspended sediment while channelisation is in progress. In October 1984, 38 measurements made at two-hourly intervals over a three-day period (Fig. 4) show peak suspended sediment concentrations of 546 mg/litre at a streamflow of only 17.52 cubic metres/second. Even at 8 cubic metres/second the concentration is 88 mg/litre. In July 1985 a much higher streamflow of 50 cubic metres/second is associated with a suspended sediment concentration of 200 mg/litre, and a streamflow of 8 cubic metres/second has an associated suspended sediment concentration of only about 40 mg/litre. These two events are not strictly comparable. The first relates to winter, the second to summer. Data from the second event, moreover, relate exclusively to the falling leg of a hydrograph when sediment loads might be expected to be smaller than for equivalent streamflow on the rising limb. The principle conclusion derived from these examples is that any flow above 10 cubic metres/second is characterised by suspended sediment concentrations above 80 mg/litre while channelisation is in progress. This is the "normal" concentration above which the maintenance of good freshwater fisheries is unlikely to be supported (Alabaster and Lloyd 1982). "Normal" conditions are not defined in terms of flow frequency by Alabaster and Lloyd. At Dunminning, a flow of 10 cubic metres/second is equalled or exceeded about 15% of the time in both pre- and post-channelisation conditions (Essery and Wilcock 1989). This is about 54 days per year. It is not, therefore, a rare flow and in flood conditions suspended sediment concentrations might be expected to be very much higher.

STREAMFLOW AND FLOODING

Prior to channelisation flooding was a major problem in the entire floodplain area between Dunminning gauge and Dunloy. Large areas of land were often flooded two or three times per year. It was to relieve this flooding and to provide a drainage benefit that the drainage scheme was undertaken. A principle area of debate at the Public Inquiry was the likely effect of channelisation on streamflow, particularly very high or low flows. An associated concern was the effect the scheme might have on the character of wetlands adjacent to the river if flooding were to be eliminated or water-tables drawn down.

As a result of the hydrological work undertaken on the upper Main since 1978 many of the above questions can now be answered (Essery and Wilcock 1989). Mean daily streamflow has been slightly augmented by channelisation, and both low flows and high flows appear to be marginally higher in post-channelisation conditions. From an ecological point of view, two effects of channelisation may be of management interest. The first concerns the effect of channelisation on water storage within the river and floodplain area above Dunminning. The second concerns the relocation of large stretches of the new river in glacial gravels previously not in extensive hydraulic contact with the River Main.

(i) Floodplain storage

Prior to channelisation, mean daily streamflow at the Dunminning gauge was frequently less than the sum of flows at the gauges on the Killagan, Cloghmills Water and River Clogh. Days when this occurs may be termed "negative flow days". Under these conditions the Glarryford catchment is an area of net water storage and prior to channelisation storage, as thus defined, occurred on 115 days per year on average. This is not surprising for the floodplain area of the Glarryford catchment is an area of extensive wetland, characterised by large tracts of raised bog and smaller, but significant, areas of fen. After channelisation the number of "negative flow" or "net storage" days is reduced to an average of 13 per year, mainly through the elimination of overbank flooding as a result of channel deepening and enlargement, and the elimination of low flow storage by the removal of meanders and pools.

The elimination of overbank flooding may prove to represent a serious impact on the area of man-modified fen which occurs on much of the floodplain in the Glarryford catchment as far north as Dunloy. Flooding represents the only obvious source of nutrients for these areas in an otherwise acidic environment, and the reduction of flood frequency from an annual, if not biannual, event to a three-year event may cause the degeneration of these fen peats into more acidic types. Areas of fen peat in Northern Ireland are small and scattered (Hammond 1981; Hamilton 1982), and represent a declining and valuable ecological resource.

(ii) The role of glacial gravels

Observations of water-table behaviour in pre- and post-channelisation conditions reveal the gravels in the floodplain to be the principal source of augmented flows in the River Main (Essery and Wilcock 1989). Water-tables in the floodplain surface deposits have not been drawn down by channelisation except within a narrow band immediately adjacent to the river channel. Nor have water-tables in the raised bog and boulder clay deposits been drawn down either (Essery and Wilcock 1989). Lowering of the new channel between Dunminning and Dunloy, however, has intersected glacial gravels along a much greater length of channel than in pre-channelisation conditions and these now make hydraulic contact with the river channel along hundreds of metres when previously gravel exposures in the bed and sides of the channel were limited. An immediate effect, perhaps beneficial in the long term, is to release gravel into the bed of the river to diversify bed material in what was formerly a uniform channel of very fine clay and organic matter. A secondary effect may be to raise pH levels at low flow but this latter effect has not been monitored.

DISCUSSION

Channelisation has major impacts on the physical conditions existing in natural rivers through its effects on channel topography, sediments and flow characteristics. Many of the changes quantified in this paper would occur on minor channelisation schemes as well as larger channelised rivers like the River Main. Most of the major changes identified in the River Main appear to lead to a more stressed environment for aquatic plants and animals and to a loss of ecological diversity in the channel and floodplain. The exact significance of some of these changes on the natural ecology of rivers is uncertain for there is as yet too little understanding of ecological processes by "physical" environmental scientists and engineers. By the same token, there is also too little understanding by ecologists of the physical processes governing flow, sediment and channel morphology and the significance of such processes in determining the total character of aquatic habitat. More dialogue between "physical" and "ecological" environmental scientists is required as well as a greater exchange of ideas between engineers and ecologists. Fortunately, this appears now to be taking place (Newson 1986; Hey 1986).

Increasingly, the concept of river corridors (Gardiner 1988) is being advanced to reconcile the interests of drainage and flood relief on the one hand with habitat conservation on the other. The river corridor concept envisages a tract of land on either side of a river in which the river is free to adjust its morphology, flow characteristics and sediment load in response to any form of induced change. The corridor also protects the aquatic system of the river from undesirable terrestrial influences of the city and/or agriculture, and is a particularly important refuge for many forms of wildlife, especially birds (Lewis and Williams 1984). River corridors also enable a river to be overdesigned so as to allow for "ecological" roughness elements (e.g. increased sinuosity, weed growth etc.) as well as mechanical roughness elements owing to channel shape and planform.

Recent calculations undertaken in the context of ecological damage resulting from pollution rather than channelisation demonstrate the high commercial value of good salmonid nursery grounds in Northern Ireland. At £85,000 per hectare this value exceeds "equivalent areas of even the best adjoining agricultural land" (Kennedy 1986, 406). Given the present policy of taking land out of agriculture, the provision of narrow river corridors perhaps represents an alternative economic land use worth detailed assessment, in view of the protection they might afford to the habitat and wildlife of rivers.

ACKNOWLEDGEMENTS

We thank the U.K. Natural Environment Research Council for funding this project between 1980 and 1988. We also thank Killian McDaid and Nigel McDowell, Department of Environmental Studies, University of Ulster, for helping to prepare the diagrams.

REFERENCES

ALABASTER, J.S. 1970 River flow and upstream movement and catch of migratory salmonids. *Journal of Fish Biology* 2, 2-13.
ALABASTER, J.S. and LLOYD, R. 1982 *Water Quality Criteria for Freshwater Fish.* London. Butterworths.

BAILEY, A.D. and BREE, T. 1981 Effect of improved land drainage on river flood flows. *Flood Studies Report: 5 years on*, 134-42. London. Institute of Civil Engineers.

BROOKES, A. 1985 River channelisation: traditional engineering methods, physical consequences and alternative practices. *Progress in Physical Geography* 9, 44-73.

BROOKES, A. 1988 *Channelized rivers: perspectives for environmental management*. Chichester. Wiley.

BRUTON, R. and CONVERY, F.J. 1982 *Land drainage policy in Ireland*. Policy Research Service No.4, Economic and Social Research Institute, Dublin.

CARLING, P.A. 1988 Channel change and sediment transport. *Regulated Rivers: Research and Management* 2 (3), 369-88.

ESSERY, C.I. and WILCOCK, D.N. 1989 The impact of channelisation on the hydrology of the upper River Main, County Antrim, Northern Ireland - a long term case study. *Regulated Rivers: Research and Management* (in press).

GARDINER, J. L. 1988 Environmentally sound river engineering - examples from the Thames catchment. *Regulated Rivers: Research and Management* 2 (3), 445-70.

GREGORY, K.J. and WALLING, D.E. 1973 *Drainage Basin, Form and Process*. London. Arnold.

HAMILTON, A.C. 1982 Peatland In J.G.Cruickshank and D.N.Wilcock (eds), *Northern Ireland: Resources and Environment*, 185-206. Belfast. The Queen's University of Belfast and New University of Ulster.

HAMMOND, R.F. 1979 *The Peatlands of Ireland*. Dublin. An Foras Taluntais.

HANNA, J.E. and WILCOCK, D.N. 1984 The prediction of mean annual flood in Northern Ireland. *Proceedings of the Institution of Civil Engineers* 77 (2) (Dec.), 429-44.

HEY, R.D. 1986 River Mechanics. *River Mechanics, Journal of the Institution of Water Engineers and Scientists* 40 (4), 139-58.

HUTTON, J.B. 1972 *River Main Drainage Scheme, Report of a Public Enquiry*. Belfast. HMSO.

KELLY, P. 1984 Ireland. In D. Baldock (ed.), *Wetland Drainage in Europe*, 48-83. Nottingham. International Institute for Environment and Development and the Institute for European Environmental Policy.

KENNEDY, G.J.A. 1986 Silage effluent pollution. *Agriculture in Northern Ireland* 60 (12), 402-6.

LEWIN, J. (ed.) 1981 *British Rivers*. London. George Allen and Unwin.

LEWIS, G. and WILLIAMS, G. 1984 *Rivers and Wildlife Handbook*. Bedford. Royal Society for the Protection of Birds.

McCARTHY, D. 1980 Impacts of drainage on fisheries. In *Impacts of drainage in Ireland*, Paper 4. Dublin. National Board for Science and Technnology.

NEWBOLD, C., PURSGLOVE, J. and HOLMES, N. 1983. *Nature Conservation and River Engineering*. Shrewsbury. Nature Conservancy Council.

NEWSON, M.D. 1986   River basin engineering - fluvial geomorphology. *Journal of the Institution of Water Engineers and Scientists* 40 (4), 307-24.

SWALES, S. 1982 Environmental effects of river channel works used in land drainage improvements. *Journal of Environmental Management* 14, 103-26.

WILCOCK, D.N. 1979 Post-war land drainage, fertilizer use and environmental impact in Northern Ireland. *Journal of Environmental Management* 8, 137-49.

WILSON, E.M. 1983 *Engineering Hydrology* (3rd edn). London and Basingstoke. Macmillan.

In: Steer, M.W. (ed.) 1991 *Irish Rivers : Biology and Management*, pp 219-229. Royal Irish Academy, Dublin.

MANAGEMENT OF THE RIVER SHANNON

Martin Quinn

Electricity Supply Board

ABSTRACT

The paper looks at various aspects of management of the River Shannon as one of our great and under-utilised natural resources, with particular emphasis on the role of the E.S.B. in regulating water flows and levels. A range of beneficial uses are identified and competing interests are discussed, as are the inevitable conflicts between some of those interests. The well-known problem of flooding and its causes are dealt with, and the author's personal views are expressed together with a clarification of the policy of the E.S.B. The Board's willingness to cooperate where possible with progressive change in the regime of control is put on record subject to consensus, or at least to wide acceptance of proposals for such change, and to statutory approval.

INTRODUCTION

The River Shannon provides enough free fuel to cater for the electricity needs of all domestic customers in the cities of Cork, Limerick and Galway combined.

This fact alone entitles me to say that my favourite river is indeed a Great River. But of course, there are many more reasons for applying the term Great to the majestic Shannon.

Circumstances often place me in a defensive role in relation to its foibles and so I am glad to have this opportunity to dwell a little on some of the more beneficial aspects of this magnificent natural national resource.

The attributes of the Shannon are indeed many, varied and impressive. Let us look at a few physical parameters (Fig. 1):

Length: 350 km; longest in the British Isles.

Flow:	5.5 thousand million tonnes of water passing Ardnacrusha each year or, if you wish, 175 tonnes of water every second of every year, on average. In terms of water management, bear in mind that this flow can vary from about 10 tonnes/sec in drought to 1,400 tonnes/sec in spate.
Catchment:	10,500 sq. km (plus 5,000 sq. km in tidal reach). Embraces all or part of 13 counties and occupies one-eighth of our total island. The main stem of the river directly adjoins 9 counties. The population of the catchment - excluding Limerick City and the estuary - is some 300,000 people, including 22 towns with population of over 1,000 and 132 settlements of under 1,000 people. The topography varies through the full spectrum of Irish landscape from barren mountainous territory through lush agricultural land to marshy lowlands and the great Irish bogs. Configuration of the basin is very flat, with a fall of only 9.3 m in 185 km through the central plains.
Lakes:	The three major lakes on the system - Allen, Ree and Derg - comprise 260 sq. km of water surface area and are supplemented by countless lesser lakes of great amenity value. Fifteen lakes are navigable; the total navigable area amounts to some 200 sq. km.
Tributaries:	The main tributaries add up to some 800 km in length, with a further 1,000 km of smaller tributaries.
Estuary:	The estuary is 100 km long and is acknowledged as one of the finest and most under-utilised deep-water ports in Europe. Available berthing draught at Moneypoint is 25 metres, capable of taking cargo vessels up to 180,000 tonnes DWT. It is complemented by a good infrastructure, including a fully serviced International Airport at Shannon - again unfortunately somewhat under-utilised.

History

In a short paper like this I could not possibly do justice to the history or folklore of the Shannon. Geologists differ on the age of the Shannon; some claim it to be 30 million years old, others have decided that it is a youngster of a mere 15 million years, born in the Miocene era. I have come across a phrase somewhere (which to me says it all) describing the great river as "liquid history flowing through the centre of our land and mirroring Ireland's historic and cultural past"; a barrier to the conquerors, a cultural divide.

BENEFICIAL USES

The benefits deriving from the Shannon can be broadly described under eight classifications:

1. *Water supplies* - for human consumption, livestock and agriculture, industrial processes, etc.

2. *Pollution control* - scavenges, dilutes, aerates and transports waste effluents - domestic, farm and factory.

3. *Fishing* - supports and sustains fishlife, game and coarse; arguably one of the finest inland fishery systems in the world, certainly unique in Europe. As a fishery owner on the Shannon, the E.S.B. plays a major role in the conservation and propagation of fish stocks.

4. *Amenity* - capacity to accommodate almost every conceivable form of water-based and allied recreational activity with virtually unlimited scope. Extensive and incomparable wetlands and nature habitat.

5. *Navigation* - 200 sq. km of navigable waters; with canal access through to the East coast and potential access to the Erne system now a real possibility if current proposals with government are found viable. There are at present over 400 hire boats on the Shannon system and a similar number of private boats - very low density for the size of the system.

6. *Energy* - the E.S.B. Power Station at Ardnacrusha substantially harnesses the available energy from the river, to the tune of 330 million units of electricity per year. This process does not degrade the quality of the water in any way, nor does it diminish the quantity; yet the savings accruing to the Board - and to the national economy - through avoided fuel imports amounts to at least £4 million annually.

7. *Drainage* - this, after all, is the primary and natural function of any river.

8. *Irrigation* - believe it or not, this can occasionally be a vital function, even in Ireland and, even more surprisingly, in the Shannon basin.

Mention of the latter two benefits regularly provokes a wry response from farming representatives, but their validity is nonetheless unquestionable.

Appendix 1, though not exhaustive, gives some idea of the range of activities and the number of disparate interests involved in the Shannon. Many of these are directly conflicting or potentially so, hence the complexity of managing the Shannon to any degree of common acceptance.

Like all great rivers (Nile, Ganges, Mississippi etc.), the Shannon can and does indeed cause serious problems in both drought and flood. Flooding of low-lying lands, particularly in the area south of Athlone but also in areas around Carrick-on-Shannon and below Banagher, is virtually an annual feature. At the other end of the scale, a dry period in spring, summer or autumn can cause some distress to boating, fishing and shore-based amenity interests.

We are almost continuously lobbied by the varied interest groups and individuals along the river; probably the most consistent and vocal are the farming, boating and fishing interests. There is a belief abroad - simplistic and grossly erroneous - that the E.S.B. has both the authority and capability, but not the will, to regulate the system in such a way as to minimise, if not eliminate, the problems of all.

Nothing could be further from the truth. Our discretionary control is no more than marginal in most circumstances; our mandate and statutory powers give us little flexibility; yet we do indeed try hard, and with some success, to train the behaviour of the river towards the greater good.

The problem of flooding, particularly below Athlone, is undoubtedly the most highly profiled problem on the Shannon. Yet the simple facts, often not appreciated, are that the natural flood flows through Athlone greatly exceed the carrying capacity of the river channel downstream, particularly the 6 km bottle-neck stretch between Banagher and Meelick (Fig. 2). In such conditions, the sluices in Athlone are long since closed and the weir is drowned, so there is no way that the E.S.B. or anybody else can hold back the torrents (Fig. 3). There seems to be little public understanding that the man-made controls in Athlone only come into effect when water levels are low; flow variations in those circumstances are of no real consequence to local farmers in any case.

Apart from this minimal control on Lough Ree, there are just two other places where we can directly influence water levels, namely Lough Allen at the upper end and Lough Derg at the lower end of the river system. I will just touch on each of these - briefly, of necessity.

The situation in Lough Derg is relatively simple. We can, and do, hold the level within a small band throughout the year except for relatively rare occasions when the network is in difficulty and we are forced to draw down below the usual range. When this happens we suffer a few slings and arrows from the boating and other interests, but they generally accept that we do not wantonly discommode them.

In flood conditions we have adequate facilities to discharge any incoming flows in a more controlled way than prevailed before 1929. Lands around the lake adversely affected by our operations have been acquired or compensated.

At Lough Allen, we fill the lake in winter - but only to pre-1934 levels - and draw down the stored water in summertime to supplement dry weather flows at Ardnacrusha. The effect of this draw-down on other parts of the river is rarely noticed, while it can be said that the vacated storage absorbs part of the late autumn/winter inflows with some resulting benefit.

And here we come across a typical example of the conflicting pressures that I have earlier referred to; the current regime suits farming interests, the fishing people are happy to have the lake half emptied but no more, while those who see boating as important urge us to keep the lake full at all times to facilitate navigation.

At this point the obvious suggestion presents itself, as it has done on numerous occasions in the past (Appendix 3) and with increasing frequency in recent times - why not bring the various interests together towards a consensus on management of the total resource? This seems like a good idea until one pauses to realise just how many and diverse those interests are, and modern-day Solomons and Jobs are in short supply!

CONCLUSIONS

To conclude, then, I will give you a few of my own personal observations on management of the Shannon in the late twentieth century.

1. The reality that certain interests are in natural conflict must be faced. Prioritised objectives should be identified and clearly stated at the highest level, based primarily on the national good but taking account, as far as is possible, of the sectional and local interests. Widespread representations should be invited and evaluated; the whole process should be fully publicised and widely discussed to develop whatever degree of consensus can be achieved. In this regard, I should tell you that the Minister of State for Finance, Mr Noel Treacy T.D., has recently announced his intention to set up a Shannon Forum. Hopefully this will be a significant milestone.

2. To my mind the concept of draining the Shannon is decreasingly likely to become a viable proposition - for financial, technical, and ecological reasons, and particularly in the Europe of today preoccupied with agricultural over-production problems. Surely we should now make a firm decision on this one way or the other and clear the decks for some real planning! We should not go on indefinitely living on aspirations unless they are realisable; it is a matter for highest-level political judgement based on inputs and expertise largely available or easily found as of now.

3. The E.S.B. is, of course, a very interested party in the management of the River Shannon. It is, however, only one of the many interested parties and in my opinion should not be perceived as the leading interest.

    I want to give a categoric assurance that the E.S.B. will not be obstructive to changes in the control regime on the upper and middle Shannon where such changes have a wide general acceptance and state approval.

    I am confident that the Board will not be found wanting in cooperation even though there are obvious financial and legal implications. The database and knowledge we have built up have been, and will continue to be, available to those who would study the problems.

224

Finally, as you will have noticed, I have of necessity made many generalised statements in the course of this paper; I am satisfied that, given the opportunity, I can substantiate and elaborate on what I have said and I welcome any such opportunity.

The challenges are obvious; the rewards great.

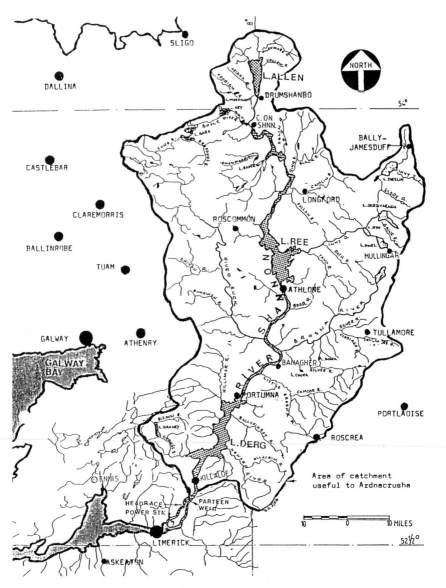

Fig. 1. Catchment area of the River Shannon. This area represents one-eighth of the total area of Ireland (one-fifth if the estuary is included).

225

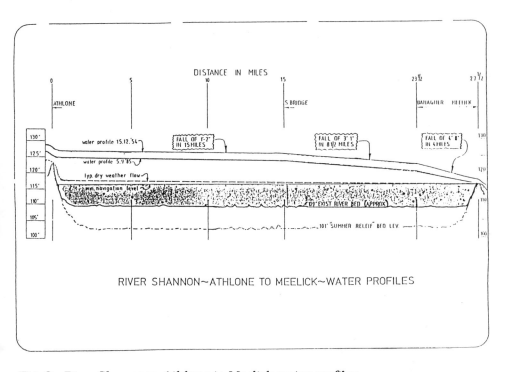

Fig. 2. River Shannon: Athlone to Meelick water profiles.

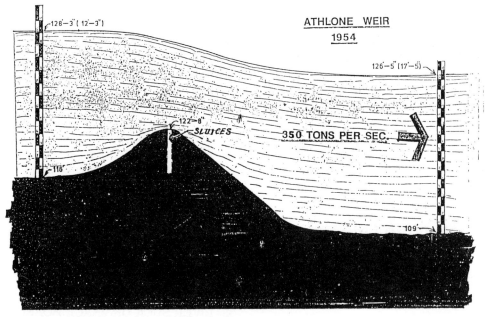

Fig. 3. Profile and velocity at the Athlone Weir in 1954.

APPENDIX 1

River Shannon

Activities	Interested parties
1 POWER GENERATION	1 E.S.B.
2 FLOOD CONTROL	2 BORD NA MONA
3 FARMING	3 O.P.W. - SHANNON NAVIGATION OFFICE
4 FISHERIES	4 O.P.W. - DRAINAGE SECTION
5 TOURISM	5 LOCAL AUTHORITIES
6 NAVIGATION	6 DEPT. OF ENVIRONMENT
7 WATER SUPPLY ABSTRACTIONS	7 FORAS FORBARTHA
8 EFFLUENT & SEWAGE DISPOSAL	8 DEPT. OF FISHERIES
9 INDUSTRIAL WATER USES	9 REGIONAL & CENTRAL FISHERIES BOARDS
10 BATHING	10 BORD FÁILTE
11 VISUAL AMENITY	11 I.D.A.
12 WATER SKIING	12 S.F.A.D.C.O.
13 REGATTAS	13 AN TAISCE
14 BOARD SAILING	14 I.A.W.I.
15 POWER BOAT RACING	15 I.F.A.
16 SAILING	16 I.C.M.S.A.
17 CABIN CRUISERS	17 MIDLANDS R.D.O.
18 PLEASURE BOATING	18 MID WEST R.D.O.
19 CRAFT HIRE OPERATIONS	19 BOAT HIRERS' ASSOCIATION
20 CANOES	20 LOCAL DEVELOPMENT ASSOCIATIONS
21 SHORE FISHING	21 WILDFOWL CONSERVANCY
22 SCUBA DIVING	22 ECOLOGY LOBBIES
23 WILDFOWLING	23 ANGLING CLUBS
24 WETLANDS ECOLOGY	24 GUN CLUBS
25 SHORE-BASED ENTERPRISES	25 WATER SPORT CLUBS
26 PICNIC SITES	26 LOCAL FARMERS' GROUPS & INDIVIDUALS
27 NATURE PARKS	27 PRIVATE & COMMERCIAL ENTERPRISE
28 HOLIDAY HOMES	28 LOCAL & NATIONAL POLITICIANS
29 FISH FARMING	29 HEALTH BOARDS
30 REED FARMING	30 CIVIL DEFENCE
	31 CO. COMMITTEES OF AGRICULTURE
	32 RIVER SUCK DRAINAGE BOARD
	33 DEPT. OF AGRICULTURE

APPENDIX 2

Note

(1) In releasing for publication the Final Report of Mr Louis E. Rydell of the United States Army Corps of Engineers on the River Shannon Flood Problem, the Government desire to invite attention to the following statement by Mr Rydell:

"The problem of Shannon River Flooding has been the subject of much study over the past 150 years. Because of the flat terrain through which the river flows, the almost imperceptible gradient of the stream with its series of lakes and connecting channels, and because of the large volume and long duration of flooding, no simple or obvious solution has heretofore been found - nor has the writer now found one."

(ABSTRACT FROM IRISH GOVERNMENT'S STATEMENT ACCOMPANYING RYDELL REPORT, 1956)

APPENDIX 3

River Shannon: Significant historical events/reports

This list is presented to illustrate the long history of management of the R. Shannon leading up to the present situation. It is not intended as a bibliographic reference source.

1755-87 : Navigation established from Killaice to Carrick-on-Shannon

1806 : Grand Canal completed

1809 : Royal Canal completed

1814 : Plassey-Errina Canal completed

   : Lough Allen Canal completed

1826 : Steam boats introduced (Dublin-Limerick)

1833 : Thos. Rhodes Report

1835 : Shannon Commissioners established

1839 : Shannon Act provided for expenditure of £600,000

1846 : Major works completed
    - dredging Killaice - Lough Allen - Lough Key
    - 7 weirs and locks (incl. Athlone and Meelick)
   : Taken over by Commissioners for Public Works (C.P.W.)

1859 : Railways had taken over most of the traffic

1860 : Ballinamore-Ballyconnell Canal (38 miles long) completed
    by C.P.W.

1863 : Bateman Report

1880 :   "   "   "   derelict

1882 : Flood sluices at Lough Allen completed
   : New channel downstream of Lough Allen completed
   : New channel downstream of Meelick completed

1887 : Allport Commission Report

1906 : Royal Commission Report

1923 : Report of Inland Waterways Commission
       (See Refs. J. Carty to M.R.D.O. Conference Athlone 1976)

1936 : New sluices at Lough Allen completed
     : Navigation into Lough Allen closed (no traffic, no objections)
     : Channel downstream of Lough Allen deepened
     : Shannon fisheries transferred from O.P.W. to E.S.B.

1940 : Drainage Commission

1950 : Navigation transferred to C.I.E.

1954 : I.A.W.I. formed

1956 : Rydell Report

1960 : All commercial traffic ceased (C.I.E. withdraw barges)
     : First hire-boat operation started

1961 : Joint Report E.S.B./O.P.W.

1963 : Report to Bord Fáilte by Dermot O'Cleary
     : O.P.W. refurbishment programme started
     : Swivel bridges replaced
     : Marinas built by O.P.W. at Carrick-on-Shannon, Banagher,
        Dromineer (Bord Fáilte-funded)
     : Mountshannon Harbour: quays at Rossmore, Meelick

1972 : Lanesboro - Tarmonbarry dredging (E.S.B.-funded)

1976 : Tourism & Recreation Study by Brady, Shipman and
        Martin (B/F & O.P.W.)
     : Athlone Conference M.R.D.O.

1978 : Canal from Battlebridge to Acres Lake opened (E.S.B. contrib.)

1985 : Conference Athlone 'Lough Ree Access' M.R.D.O.
     : Canals passed from C.I.E. to O.P.W.

In: Steer, M.W. (ed.) 1991 *Irish Rivers : Biology and Management,* pp 231-253. Royal Irish Academy, Dublin.

WATER QUALITY MANAGEMENT PLANS FOR IRISH RIVERS

Paul F. Toner

Environmental Research Unit, Dun Laoghaire

ABSTRACT

Increased use of water for various purposes has created a need for systematic management of this basic resource. Water quality management planning requires information and policy decisions on a range of matters but particularly beneficial uses, water quality criteria and standards, and on waste sources and their control. In the late 1970s and early 1980s, a series of draft plans was prepared by An Foras Forbartha, on behalf of groups of local authorities, for a number of river systems. The main policies in these draft plans were the formulation of water quality standards based on the water quality criteria for salmonid fisheries and the control of waste discharges by the environmental quality objective approach. The draft plans mainly address the control of pollution by organic (BOD) wastes from point sources. Further development of the plans must allow for the control of non-point source pollution and of eutrophication in rivers.

INTRODUCTION

Water is one of the basic natural resources underpinning man's social and economic progress. Over the last two centuries, a tremendous increase in the human population and associated developments in industry and agriculture have led to great pressures on water resources in many countries. These pressures relate not only to the availability of water but also to water quality. While, on the one hand, enhanced living standards require ever greater usage of clean water, on the other hand very large quantities of water-borne wastes are generated by domestic, industrial and agricultural activities. Since in most cases these wastes, or their residue following treatment, must be disposed of by discharge to surface or groundwaters, a potential problem is created if the introduced wastes degrade the quality of the recipient waters below the level required to sustain fisheries, abstraction for domestic purposes or any other beneficial uses.

Initial attempts to resolve or minimise these conflicts led to the development of methods for the large-scale treatment of water and waste waters. These advances alleviated the most serious problems involving human health and gross pollution of rivers and other waters. However, in more recent times the need for comprehensive rather than piecemeal management of water resources has become clear. This new approach has occasioned a great deal of research into the natural characteristics of aquatic ecosystems and their response to the introduction of wastes as well as into waste treatment technology. The developed countries now spend very large sums on the protection and conservation of water resources. The need for similar measures in many of the Third World countries is even more pressing in view of the fundamental problems for agriculture and the prevention of disease which lack of water or of clean water presents in such cases. The recent UN-organised International Drinking Water Supplies and Sanitation Decade has attempted to alleviate some of these problems.

Use of rivers as sources of water supply and as the means of disposing of wastes, particularly those from urban areas, is undoubtedly the most important consideration from the economic and social point of view. However, fisheries, recreational and amenity uses and the preservation of natural flora and fauna also have strong claims for consideration. These aspects have come much more to the fore in recent decades owing to an increased awareness of environmental degradation and a general interest in wildlife and its conservation. It is the objective of water quality management to reconcile the often conflicting demands of all these aspects and to ensure that the quality of water is sufficient to accommodate present and potential beneficial uses wherever and whenever required. It is clear that the nearer water quality conditions are kept to the natural situation the greater will be the chances of achieving this objective.

In Ireland, interest in and the need for systematic management of water quality are of relatively recent arrival. This situation reflects the much smaller pressures on water resources here compared to many other countries owing to the small population, based mainly in coastal areas, and the still only moderately developed industrial and agricultural base. While there have been significant increases in population and in industrial and agricultural activity over the last three decades, and a measurable degradation of water in some areas (Toner *et al.* 1986), the overall situation is good and compares very favourably with that in most other European countries (Toner 1987). However, this situation creates its own problems in terms of maintaining a generally high level of water quality in rivers and the consequent need for a relatively stringent approach to pollution control.

The basis for a more systematic approach to the problem than that possible under previous legislation was provided by the bringing into force of the Local Government (Water Pollution) Act, 1977 (now the subject of an Amendment Bill). Among its more important provisions, the 1977 Act allowed for the licensing of trade discharges and the preparation of water quality management plans. In the latter case, there is a provision for groups of local authorities to jointly prepare plans for specific waters, thus circumventing the problem caused by the fact that functional boundaries of the authorities do not coincide, in many cases, with the boundaries of river catchments.

In the late 1970s and early 1980s, An Foras Forbartha (AFF) was commissioned by groups of local authorities to prepare draft water quality management plans for river catchments in several areas of the country, including the Suir, Nore, Barrow and Slaney in the south-east and the upper and lower Shannon catchment. The proposals and discussions set out below are largely based on these draft plans of which the writer was joint author. However, before considering the details of these draft plans it is considered useful to discuss some of the general issues involved in water quality management planning.

GENERAL CONSIDERATIONS

The main objective of a water quality management plan is to ensure that the condition of the water body under consideration is suitable to support present or potential beneficial uses. The basis of the plan consists, therefore, of a set of policy decisions on the general measures to be adopted to meet these objectives. The measures adopted relate, firstly, to the prevention of pollution and, secondly, to the restoration of those waters where quality is presently unsatisfactory. Before these policy decisions can be made, it is necessary to obtain a wide range of data and information on the river system and related matters, some of which may already be on record and some of which may require the carrying out of special surveys and investigations.

Information and data requirements

The more important needs in relation to information and data in the formulation of water quality management plans are summarised in Table 1. The table lists the most critical requirements in each of the general areas. These are commented on further below.

Present water quality status. In many cases useful information is likely to be available from existing survey programmes. However, additional investigations will usually be required in order to e.g. define more clearly the impact of waste discharges under low flow or other critical conditions, to establish the position with regard to less common pollutants which are not the subject of routine measurement, or to establish in some detail the relationships between particular waste inputs and water quality. The database so acquired would be expected to constitute a baseline for future reference and would, in particular, identify the presently polluted reaches. A water quality map, based on these data, is a useful inclusion in a plan and its regular updating will give a ready impression of the impact of any remedial measures undertaken on foot of the plan.

Table 1. Information and data requirements for water quality management plans.

General requirement	Critical requirement
Present water quality waste status of the river system.	Condition of reaches in receipt of discharges. Identification of presently polluted reaches.
Present and potential beneficial uses.	Details of those uses most likely to be adversely affected by pollution, e.g. fisheries and water abstractions. Present impact of pollution on such uses.
Hydrological aspects.	Magnitude of low flows in rivers. Flushing rates for lakes. Dispersal characteristics in tidal waters.
Waste generation and disposal.	Characteristics of major point sources of waste (sewerage systems and industrial plants). Non-point waste sources in catchment, particularly locations where waste, e.g. manure slurry, is stored and adequacy of facilities used.
Water quality criteria.	Most up-to-date information on the water quality criteria applicable to the most pollution-sensitive uses, e.g. fisheries.

Present and potential beneficial uses. A considerable effort of documentation may be required to cover all the uses and activities which depend on the river system. Clearly, those uses which are affected or likely to be affected by loss of water quality require particular attention. In such cases, the need for detailed information may necessitate special investigations, e.g. in relation to the nature of the fish stocks. In other cases, careful consideration may be needed in order to assess future developments, e.g. an increased level of water abstraction. It is desirable that a wide range of opinion, both official and non-official, be sought on this aspect of the plan so that all possible uses and needs are catered for by the pollution control measures finally adopted.

Hydrological aspects. The dilution of wastes discharged directly to the river system is a critical consideration in formulating the pollution prevention measures incorporated in the plan. In rivers, the chief constraint on the discharge of waste is the magnitude of the low flows and, in most cases, the level of waste treatment required is dictated by the dilution available during such flows. This constraint can be avoided where facilities are available to store waste during periods of low flow. In general, storage is not practicable because of lack of space at a convenient point and for aesthetic and public health reasons. In some cases, the natural low flows may be augmented by stored water from upstream impoundments or from groundwaters. Such measures, if taken only for water pollution control purposes, are likely to be difficult to justify on economic grounds.

In general, therefore, the permissible waste discharge at a particular location is determined by the low flow dilution capacity. However, use of the lowest measured or recorded flow is not considered justifiable, at least in the case of non-toxic wastes, since the expected return period of such a flow is likely to be of the order of 20 to 30 years. Generally a somewhat higher flow is taken, e.g. the 95 percentile flow (that is the flow expected to be exceeded for 95 per cent of the time on a long-term basis).

Reliable information on the low flow statistic chosen and on other flows generally depends on the availability of long-term records from suitably located gauging stations. Where such records are not available, estimates have to be made on the basis of records from other locations in the catchment or in adjacent catchments. This procedure is clearly prone to error; the potential problems can be partially circumvented by carrying out flow measurements at times of drought. In addition, allowances must be made, where appropriate, for the effects of abstractions on low flows, in particular those cases where the abstracted water is not returned to the river.

In lakes, the flushing rate and mean depth have a strong influence on the response to the excess loading of nutrients owing to waste inputs. Estimates of these factors require, in turn, data on the flows in the feeder streams and on the morphological characteristics of the lake. In many cases, therefore, field investigations will be required to obtain the necessary information. Estuaries represent an extension of the riverine situation but with the added complications caused by tidal oscillations. Estimation of the dilution and dispersal rate of wastes discharged to estuaries is likely to require mathematical modelling backed up by a relatively elaborate field investigation programme.

Waste generation and disposal. The discharge of waste to rivers under properly controlled conditions is clearly of great beneficial use to the community. However, because of its potential conflict with other uses, this aspect requires separate consideration. Information on waste generation and discharge is more likely to be already available and more easy to obtain by measurement in the case of point sources (e.g. sewerage systems) than it is in the case of non-point sources (e.g. agricultural activities). For point sources it is necessary to know the total pollutant load generated in each case, the level and efficacy of treatment, if any, and the nature of the discharged effluent. Information on seasonal variations in waste generation, as occur in many industrial operations, is also required in order to make a proper assessment of the situation. If the data necessary to calculate waste loads, viz. discharge rates and pollutant concentrations, are not available or sparse, it may be necessary to undertake detailed sampling and analysis of the effluent as well as volumetric measurements. In some cases, particularly domestic wastes, reasonable estimates can be made based on generally accepted unit waste loads, e.g. 54 g BOD per person per day in sewage. This approach is not so easy in the case of industrial wastes except where the plant under consideration has a close parallel elsewhere for which waste generation per unit of production is well documented.

In the case of non-point source wastes the most important aspect to be covered is the adequacy of storage as deficiencies in this area have a potential to cause gross pollution of ground and surface waters. The locations at which significant quantities of waste are stored must be documented and the pollution risk assessed in each case on the basis of the storage facilities and how these are operated and maintained. The storage of toxic or otherwise hazardous waste in the catchment requires special attention, particularly where this constitutes a threat to water supply sources. Information is also required on the ultimate disposal of the stored waste. This relates mainly to the land-spreading of farm wastes such as manure slurries. Unless spreading is undertaken with due consideration for factors such as soil type and climatic conditions, pollution of surfaces and groundwaters is likely to occur owing to direct run-off of the spread waste or to excessive leaching. Full information on non-point waste sources and their potential to cause pollution is unlikely to be on record and special investigations may be needed to meet data requirements in this area.

The policy considerations in the plan also require some estimate of future waste loads as these will have a bearing on the decisions to be made on waste treatment. Such estimates can be calculated on the basis of population and industrial projections. In the latter case, particularly, estimates are likely to be largely speculative and the same status may attach to any estimates of future waste load generation in the agricultural sector. The best that might be attempted is to calculate a range of estimates for each sector and to assume an overall increase of waste generation which is likely to err on the high rather than low side.

Water quality criteria. A voluminous body of information now exists on the relationship between the concentrations of specific pollutants in water and their effects on various uses. Typical examples are the publications of the US Environmental Protection Agency (Train 1979) and the European Inland Fisheries Advisory Commission (Alabaster and Lloyd 1980). The water quality criteria provide a basis for the setting of water quality standards. Any standards adopted must ensure that the quality of the water is adequate to satisfy all uses. Thus, for any particular parameter, the water quality standard must be based on the criterion applicable to the most sensitive use.

In general, the most stringent criteria apply in the case of fisheries, particularly the salmonid type, and water quality standards are often based on such criteria. Thus, in addition to those required by fisheries, the water quality conditions required by most other uses are assured. There are some exceptions to this, e.g. the criterion for nitrate levels in water for abstraction is considerably more stringent than that for fisheries, while the restriction on bacterial contamination required in bathing waters is not of great relevance in respect of fish. In general, however, water quality standards based on fishery requirements provide the best overall protection for beneficial uses and, for this reason, promotion of fisheries is usually a basic objective of water quality management.

Many countries have legally enforceable standards for water quality. In Ireland water quality standards are provided for in the Local Government (Water Pollution) Act, 1977, but this provision has not yet been activated. However, since the mid-1970s several directives issued by the EC have dealt with the quality of waters for various uses, e.g. abstraction of water for public supply, bathing and freshwater fisheries (Council of the European Communities 1975; 1976; 1978). Limits set in these directives for potential pollutants of water are legally binding in certain circumstances specified in the directives, e.g. designated rivers in the case of the Freshwater Fish Directive; where the waters covered by the directives do not presently meet the limits specified, improvements must be undertaken to ensure compliance within a specified period. While the directives are legally binding only in the defined circumstances and waters, the pollutant limits included are widely referred to as a general guideline for the setting of water quality standards.

Policy decisions

The basis of a water quality management plan, as mentioned above, is a set of policy decisions on the approach to water pollution control and the undertaking of remedial measures where necessary. These decisions must be made in the light of the data and information described in the foregoing section. The main policy decisions to be made concern the setting of water quality standards and the general and specific controls for point and non-point waste sources. In addition, there is a need to decide on the priorities of a remedial programme for presently polluted reaches.

Water quality standards. These are determined, as pointed out above, by the water quality criteria applicable to the most sensitive of the beneficial uses to be protected. In general, water quality standards are identical, in their restrictions, to the corresponding criteria. However, in some cases, e.g. where the criterion is based on limited data or on acute effects only, a more stringent limit may be adopted for the standard. In other cases, particularly for non-toxic pollutants, the setting of a limit for the standard which is less stringent than the criterion may be justified for economic or other reasons. There may be a case for altering the standard in different parts of the river system in order to take account of natural factors, such as hardness, which may have an ameliorating or an intensifying effect on the impact of specific pollutants. This is recognised, for example, in the EC Freshwater Fish Directive in respect of the metals zinc and copper, for which the limits to apply in soft waters are more stringent than those applicable in hard waters. It has been argued recently that a rigid adoption of the 'no-effect' value of the water quality criterion as the standard may be a too pessimistic extrapolation of laboratory data since such values reflect a worst case situation which is not often realistic in practice (Lee *et al.* 1982).

Besides the consideration of natural factors in setting standards there may be a case for differentiating between the water quality requirements of different parts of the river system on the basis of beneficial uses. This often arises in the case of fisheries where, for natural reasons or because of what may or may not be practicable in restoring polluted waters, some parts of the river system are designated as suitable for salmonid fish and other parts only for the less pollution-sensitive coarse fish. Differentiation in respect of water quality standards might also apply if some stretches, historically affected by gross pollution, are not considered capable of full restoration and where the water quality objective is thus mainly designed to prevent public nuisance. In contrast, especially stringent conditions may be necessary in stretches from which water is abstracted for public supply. However, in any such zonation of the river system, on the basis of water quality requirements, account must be taken of the needs of downstream reaches, e.g. of the main channel below a presently grossly polluted tributary.

In setting water quality standards account must be taken also of the methods whereby compliance is to be assessed. In theory, the water quality limits should be respected at all times and under all conditions. However, in practice, when compliance is assessed on the basis of a relatively small number of samples taken over a year or other fixed period, some allowance must be made for small variations in the test procedures or in river conditions which, while of little significance for the designated beneficial uses, would imply a breach of a rigidly interpreted limit. It is current practice, therefore, to formulate the standard so that compliance with the water quality limits is required for a period somewhat less than the total period in question or for a proportion only of the samples taken. The commonest such proportion for sampling period or sample numbers used is 95 per cent.

Compliance in a specified proportion of samples is a relatively simple approach and is used in the limits set in the EC directives. Time-based compliance (95 percentile limit) requires assessment by appropriate statistical techniques and the sampling programme needs careful planning to allow the valid operation of such an approach. In both cases, the numbers of samples taken must be high enough to be reasonably representative of the variations likely to occur. The EC Freshwater Fish Directive requires a minimum monthly sampling frequency. However, proper assessment of compliance with the time-based percentile limit is likely to require a sampling frequency of at least twice per month. For sample numbers less than ten, there probably should not be any breaches of the limit.

Control of waste sources. There are two main aspects to be considered under this heading, viz. the regulation of effluents from point sources of waste and the minimising of waste losses from non-point sources. In general, the former is likely to be the primary concern and also easier to address in a systematic manner.

Regulation of effluents from point sources may be approached in two ways, viz. by a uniform system of emission standards or by the environmental quality objective approach. Uniform emission standards require that certain basic limitations on pollutant levels in effluents or on the amounts of pollutants generated per unit of production be adhered to irrespective of local circumstances. This approach is particularly relevant in the case of certain toxic pollutants which tend to accumulate in animal or plant tissue and for which a strong case can be made, therefore, to limit the amounts entering the biosphere. The uniform emission approach has been incorporated in the EC directives on the discharges of dangerous substances although allowance is made for combining this with the environmental objective approach, where appropriate.

In the case of biodegradable wastes, such as sewage, the case for uniform emission standards is not as convincing as it is for toxic pollutants. Because of their biodegradability a certain quantity of such wastes, dependent on local circumstances, can be assimilated without breaching the water quality standards. Use of this waste assimilation capacity (WAC) in a controlled manner is referred to as the environmental quality objective approach. The WAC is mainly a product of dilution capacity and upstream quality; this gives a quantity of waste which can be added to the river without breaching downstream limits. This quantity in turn indicates the level of control required for effluents.

The main consideration in calculating WAC is dilution, especially the dilution available under low flow conditions. As pointed out above, the minimum or drought flow is likely to be unduly restrictive because of its relatively low occurrence and a somewhat higher flow is usually considered, e.g. the 95 percentile. In most situations, therefore, the level of control required for effluents is determined by the WAC available at the low flow condition selected. Clearly, this implies that the level of waste treatment required will be much greater than necessary at higher flows. There has been much discussion in recent years, especially in the US (e.g. Downing and Sessions 1985), of a more flexible approach to effluent control, e.g. seasonally adjusted limits based on the varying WAC, in order to lessen the impact of this restriction. However, there may be practicable difficulties in such an approach, e.g. in respect of treatment plant operation, and also a need for close policing of effluent quality.

The specific controls needed for individual effluents or groups of effluents are assessed on the basis of their present impact under the limiting conditions proposed. This is often done by means of a simple mass-balance equation using a parameter, such as BOD, to characterise the pollutant. While use of the mass-balance equation to obtain a precise measure of the water quality condition downstream of a discharge is likely to be misleading, a flaw which can be circumvented by appropriate statistical techniques if sufficient information is available (Warn and Brew 1980), it can give a useful indication of the scale of treatment required to ensure compliance with the water quality standards.

In the case of sewage and other biodegradable wastes it may be necessary, however, to undertake a more detailed investigation of the possible impact on receiving waters, in particular on the dissolved oxygen regime. The self-purification process by which biodegradable wastes are degraded in receiving waters constitutes a potential threat to dissolved oxygen levels, particularly where reaeration rates are relatively slow. Prediction of dissolved oxygen variations in such situations requires the use of mathematical models allied with field investigations to measure key factors such as reaeration rates, time of travel and the dissolved oxygen uptake and production rates of the benthic organisms. The latter are of particular importance in shallow rivers and often account for most of the deoxygenation which occurs below discharges of biodegradable wastes.

In most lakes the main concern is the stimulation of planktonic algal growth by the phosphorus in waste discharges. Relationships have been established between phosphorus input and algal growth in lakes (e.g. OECD 1982). These relationships give a basis for assessing the need for controls on phosphorus inputs. However, where waste inputs are likely to constitute a significant proportion of the total phosphorus loading on a lake, it is preferable to divert the discharges to the lake outflow, if possible, or to remove as much as possible of the nutrient before discharge. This approach is especially desirable in the case of those lakes supporting salmonid fisheries or water abstraction, uses which are best suited by low or only moderate levels of algal growth.

The control of non-point sources of waste requires a different approach to that for point sources. In the latter case a certain quantity of waste is permitted to be discharged to receiving waters. In the case of the non-point sources, there is a *de jure* expectation that the amount of waste entering waters will be zero allied to a *de facto* approach in which carry-over of wastes to waters is to be minimised. Compared to the situation for point sources, the control of non-point sources, particularly agricultural activities, is inherently more difficult. This is due to the large number of locations, many of them often remote, at which the waste is generated, the special requirements for interception and storage and the mode of final disposal which, in many cases, means application to large areas of land. The situation creates problems for the formulation and policing of regulations and for the monitoring of possible impact on vulnerable waters.

Interception and storage of waste is of critical importance since failure in this area is likely to give rise to very serious pollution of surface or groundwaters. Satisfactory arrangements for the interception and storage of waste are generally sought under planning regulations; this approach may be supplemented by requirements attached to grants for farm development which specify the provision of adequate waste storage facilities. In Ireland, there has been some concern regarding the exemption of certain levels of farm development from the planning process because of the implications for waste control. Amending regulations issued in 1984 have considerably reduced the scope for exemption and it is likely that most of the significant new or expanded farming development now requires planning approval.

Ultimate disposal of the waste by land-spreading presents its own problems as regulation of this activity is usually not possible within the context of basic water pollution control legislation. Since there is no direct discharge to waters, the wastes cannot be regarded as effluents. In general, minimising pollution from land-spread wastes requires that application rates appropriate to the type of land and climatic conditions prevailing in each case be adhered to and that spreading be avoided during unfavourable periods, e.g. during heavy rain, frost or snow or when plant growth is negligible. These restrictions have an influence on storage requirements which must be sufficient to accommodate waste generated during periods when land-spreading is likely to cause pollution. Specification of the correct measures to be adopted in land application of wastes may be made in planning permissions or, in certain circumstances, under water pollution control legislation. However, even more so than in the case of waste storage, proper policing of any restrictions laid down is likely to be difficult and establishing the connection between cause and effect, where water pollution arises from land-spreading of wastes, practicably impossible in many cases.

Restoration of polluted reaches. Of all the elements of a water quality management plan, the restoration of polluted waters is likely to have the greatest public impact and for many will be the touchstone of the plan's effectiveness. Ideally, the nature of a restoration programme should be guided by a cost-benefit analysis. Thus, the need to improve waters from which abstraction for public supply is made may be judged to be of greater importance to the community than reduction of pollution in other waters where fisheries are the main concern. Again, improvements to polluted sections of the main channel may be assessed as taking precedence over polluted tributaries.

However, it is unlikely that political and other considerations would allow a straightforward analysis of the situation in this manner (even if the necessary information were available). This is particularly the case in situations, e.g. in Ireland, where the overall water quality position is good and where there is a consequently high expectation that improvements should be made in all cases. Satisfactory resolution of the situation, where available resources are scarce, presents major problems for regulatory agencies, although there is likely to be a general consensus that the most seriously polluted waters merit immediate attention.

Definition of the measures required to restore particular waters requires a careful inventory of all waste sources likely to be involved in the pollution observed and an evaluation of their relative importance. The most tractable situations are those where only one or two point sources are involved and where, therefore, the necessary improvements in waste control can be readily identified. Deciding on the most effective approach to water pollution control may be more difficult where a large number of waste sources contribute to the problem, particularly if non-point waste sources are also involved. In such cases it may not be easy to decide whether or not the relatively straightforward measures to deal with point sources will be sufficient to bring about the required improvements; this often arises in cases of lake eutrophication and may require detailed studies of nutrient inputs before effective pollution control strategies can be defined with confidence.

The extent of improvement to be achieved in any particular case also needs consideration. For grossly polluted waters it may not be practicable to undertake improvements which would make these suitable for the more quality-sensitive uses, e.g. game fisheries; the main objective might be to prevent such waters creating a public nuisance by ensuring that anaerobic conditions do not occur. In the case of less seriously polluted waters there is likely to be a generally justifiable objective of making these suitable to support most beneficial uses. In general, the main objective is likely to be the restoration of conditions suitable for fisheries, at least the relatively tolerant coarse fish. The presence of viable fisheries is not only an assurance to water resource managers that quality conditions are generally suitable for most uses but is likely to be the main criterion by which the public will judge the effectiveness of pollution control measures.

PLANS FOR IRISH RIVERS

General situation

The main objective of water quality management planning for Irish river systems is the maintenance of the present good conditions. These good conditions permit the ubiquitous occurrence of salmonid fish which is one of the chief characteristics of Irish inland waters. A further point of note bearing on water quality requirements is the fact that the bulk (~80 per cent) of the public water supply is abstracted from rivers, lakes and other surface water bodies which are more likely to experience direct pollution than are groundwaters. This is the opposite of the situation in many other European countries where most of the supply comes from groundwaters.

With regard to waste generation in Ireland, not only are the overall quantities generated relatively small but the direct discharge of point source waste to inland waters is only 20% of the total of the discharge from such sources (Water Pollution Advisory Council 1983). Farm wastes have assumed a greatly increased potential to cause pollution over the last two decades owing particularly to the adoption of intensive livestock rearing practices and the use of grass silage for fodder. However, the overall impact of these wastes, which greatly exceed those generated in the domestic and industrial areas, is relatively small owing to the fact that the bulk is produced by grazing animals and to the disposal of the remainder by land-spreading. This assessment must be qualified by the recognition that many of the most serious incidents of water pollution in recent years have been caused by the improper disposal of farm wastes.

The great bulk of the waste generated in Ireland is of the organic bio-degradable type. Because of the lack of heavy industry there is relatively little occurrence of waste containing toxic substances, such as metals, although the potential for contamination with synthetic organic substances may be increasing owing to the growth of pharmaceutical and allied chemical industries. Thus, the main pollution effects likely to occur in Irish rivers and other waters are deoxygenation and the growth of saprobic organisms, e.g. bacterial slimes (BOD effects) and the stimulation of algal and weed growth with possible secondary deoxygenation (eutrophication effects). These forms of pollution are well understood and the technological and other measures needed to control them are readily available.

The generally good situation in Irish rivers created little interest in the systematic collection of water quality data in the period up to the 1970s. Regular national surveys commenced in the early 1970s with the formation of the Water Resources Division in AFF. Following the implementation of the Local Government (Water Pollution) Act, 1977, many of the local authorities set up their own laboratories to monitor waters and licensed effluents. In addition, three regional laboratories were set up in Castlebar, Kilkenny and Monaghan to undertake water quality monitoring and allied tasks for groups of local authorities in the west, south-east and north-east.

In the late 1970s the local authorities, with the assistance of AFF, carried out a survey of abstractions and waste discharges. The survey covered both public and private abstractions and discharges. However, information on waste characteristics was mainly confined to volumetric discharge rates. Also in the 1970s, the local authorities started to establish hydrometric schemes to generate river flow data needed for water resources development and pollution control. The existing gauges, most of which were installed in connection with flood control and drainage schemes, were not fully reliable at the lower end of the flow range where the main interest lies in relation to water pollution control.

Background to plans drafted by AFF

The information available on water resources increased greatly, therefore, in the 1970s, both in relation to quantitative and qualitative aspects. However, at the time that AFF commenced drafting the first water quality management plans for rivers in the south-east, the databases were still relatively small and information was deficient in a number of areas. In particular, very little information was available on non-point sources and their effects, although the impact of such wastes at that time in the rivers concerned did not seem large. Thus, the draft plans mainly addressed the control of pollution from point sources.

It was decided to concentrate mainly on the BOD impact of the wastes as this was their most easily characterised fraction. In addition, the most serious pollution occurring in the rivers under study was primarily a BOD type effect, although eutrophication influences were also apparent. However, the latter aspect was regarded as requiring further research in respect of possible control measures, particularly in rivers. The topic was not addressed therefore in these first draft plans with the exception of lake waters in catchments of rivers, such as the Shannon, which were dealt with later.

The initial plans drafted by AFF covered the freshwater reaches of the Suir, Nore and Barrow, and the freshwater and estuarine reaches of the Slaney. Subsequently, AFF was commissioned to prepare a draft plan for the joint estuary of the Suir, Nore and Barrow. Following this first group, further draft plans were prepared for the Liffey, the Upper and Lower Shannon, the Shannon estuary and for County Cavan. The main features of these draft plans are outlined below.

General features of plans drafted by AFF

General methodology for plans. The methodology adopted in the preparation of the draft plans consisted of a number of stages, as follows.

- The available information on water quality and quantity was compiled and analysed to determine e.g. stretches where water quality improvements were required and the dilution available for waste discharges under low flow conditions at particular locations of interest. The databases examined were mainly those generated by AFF and the Regional Water Laboratories on water quality and the hydrometric records from OPW gauging stations.
- The available information on those beneficial uses of the river system sensitive to water quality impairment was obtained.
- The extent of the river system to be formally included in the plan was defined and depicted on a map.
- Relevant documentation was consulted in regard of the water quality criteria appropriate to the beneficial uses identified; the latter formed the basis for the specification of the water quality standards to be incorporated in the plan. In carrying out this study, particular note was taken of the limits set out in relevant EC directives and of the Water Quality Guidelines issued by the Technical Committee on Effluent and Water Quality Standards of the DoE.
- Estimates were made of the waste assimilation capacities at the locations of interest in the river system. The waste assimilation capacity is defined as the maximum amount of waste which can be discharged under specified low flow conditions without exceeding the limits set by the water quality standards.
- Estimates were made of the existing waste loads at specific locations and of the loads likely to occur at some time in the future, generally twenty years hence.
- The impact of present and future waste discharges under the specified low flow conditions was estimated and the capacity for additional waste assimilation, where appropriate, was calculated.

Table 2. Main water quality standards incorporated in the Draft Water Quality Management Plans prepared by AFF.

Parameter	Units	Limits	
Dissolved oxygen[1]	$mgl^{-1}O_2$	50 percentile	: 9
		95 percentile	: 6
		Minimum	: 4
BOD[1]	$mgl^{-1}O_2$	50 percentile	: 3
		95 percentile	: 5
Ammonia (total)	$mgl^{-1}N$	95 percentile	: 0.5
Ammonia (un-ionised)	$mgl^{-1}N$	95 percentile	: 0.02
Oxidised nitrogen	$mgl^{-1}N$	95 percentile	: 11[2]
Phosphates (rivers)	$mgl^{-1}P$	50 percentile	: 0.1
		95 percentile	: 0.2
Total phosphorus[1] (lakes)	$mgl^{-1}P$	Annual mean	: 0.05

[1]In coarse fish waters, the following limits were specified:

Dissolved oxygen	$mgl^{-1}O_2$	50 percentile	: 6
		95 percentile	: 3
		Minimum	: 2
BOD	$mgl^{-1}O_2$	50 percentile	: 4
		95 percentile	: 6
Total phosphorus (lakes)	$mgl^{-1}P$	Annual mean	: 0.08

(The total phosphorus limits are based on the OECD tropic classification scheme. Data for the clearer-water Irish lakes suggest that these limits are too lax in such cases. For these clear-water lakes, e.g. Loughs Ennell and Sheelin, it is probable that the annual mean concentration of total phosphorus should not exceed 0.025 $mgl^{-1}P$. The limits for phosphorus in both rivers and lakes must be regarded as very tentative and further cause/effect data are required to make a proper assessment of the situation.)

[2] This is the generally accepted limit for abstraction for public supply.

- The measures required to improve conditions at those locations where the waste assimilation capacity was overloaded or where the assimilation capacity was limited were assessed. A similar analysis was carried out for the projected future waste loadings.
- The priorities for investment in public and private waste treatment facilities were identified.

The extent to which this methodology could be implemented in full was constrained to a greater or lesser extent in the case of individual river systems and locations by the amount of information and data available. In particular, data needs could not be fully met in many cases with regard to flow statistics and to waste loads, especially industrial waste loads. In such cases estimates had to be made based, e.g., in the case of river flow statistics, on comparisons with other locations where information was available.

Water quality standards. In nearly all waters dealt with, the water quality standards were based on the water quality criteria applicable to salmonid fisheries since these are more stringent, for most parameters, than those applicable to other uses. The only exceptions were some waters in Co. Cavan where salmonids are of little importance compared to coarse fish; in these cases water quality standards based on the criteria appropriate to coarse fisheries were defined. Specific standards are incorporated in the plans for five parameters, viz. dissolved oxygen, biochemical oxygen demand (BOD), ammonia, oxidised nitrogen (nitrate + nitrite) and phosphate (Table 2). These parameters give a measure of the level of pollution due to biodegradable organic wastes, the types which form the bulk of those likely to be encountered in Ireland. In the case of estuaries, specific consideration was also given to microbiological standards in relation, particularly, to the quality requirements in shellfish and bathing waters.

Waste assimilation capacity (WAC). The policy advocated for the control of waste discharges is based on the environmental quality objective approach, i.e. the receiving water is deemed to have a waste assimilation capacity. For rivers and streams the waste assimilation capacity was defined as the waste load (as BOD) which can be discharged under 95 percentile flow conditions without the downstream BOD concentration exceeding 4 mg l^{-1}. In estuarine waters, WAC for organic wastes was calculated with reference to the dilution available under conditions of neap tides and 95 percentile flows in inflowing rivers and to a maximum BOD concentration of 4 mg l^{-1}. In the case of such waters, the draft plans incorporate mathematical models of the waste dispersal characteristics in order to determine the dilution capacity under critical conditions.

In those catchments where lakes occur, e.g. the Upper and Lower Shannon catchments, tentative measures to control phosphorus inputs are incorporated in the draft plans. The need for phosphorus control recognises the primary role of this element in the promotion of excessive plant growth due to eutrophication. The relationships established in the OECD study (OECD 1982) between phosphorus input and planktonic algal growth (measured as chlorophyll) were examined in the draft plans as to their suitability for assessing phosphorus assimilation capacity in Irish lakes. In view of existing data for phosphorus input and algal growth in such lakes, these relationships are not likely to offer any more than a very general guideline on necessary control measures as the predictions based on the relationships have a very large margin of error. As an alternative approach, it was suggested in the draft plans for the Shannon catchments that the input of phosphorus from waste sources should no more than double the bioavailable phosphorus input from natural sources. It was further recommended that the latter be calculated as only 10% of the total phosphorus input from natural sources, using a typical 'catchment input' rate of around 25 mg P m^{-2} yr^{-1} for this input. This approach, therefore, allows for the much greater bioavailability of phosphorus in wastes compared to that in natural sources. Calculations indicate that it is likely to be a quite stringent control strategy for phosphorus inputs.

In the case of potentially toxic pollutants, the plans indicate that waste assimilation capacity should be calculated using the minimum rather than the 95 percentile flow. The plans give detailed listings of water quality criteria for such pollutants with recommendations as to the limits which should be adopted in the plans.

Assessment of the impact of waste discharges. In assessing the impact of waste discharges in rivers and streams a simple mass-balance equation was used to calculate the theoretical BOD concentration which would arise below the discharge point under 95 percentile flow conditions. This concentration, together with the residual waste assimilation capacity (if any), was used as an index of the need for improvements. Improvements were considered necessary if the calculated downstream concentration exceeded 4 mg l^{-1} or if the residual waste assimilation capacity was limited, in view particularly of likely increases in waste generation. It is important to note that the calculated downstream concentration of BOD is not a forecast of the 95 percentile value which, as pointed out above, requires detailed statistical assessment. The assessment of the impact of waste discharges also took account of the available water quality data. In general there was little conflict between the calculated condition and that observed.

Waste treatment needs. The basic policy of the draft plans in respect of waste treatment is that there should be a minimum requirement for primary treatment in all cases, i.e. there should be no untreated discharges. In relation to specific waste treatment needs, only three possible strategies were considered, viz. primary, including preliminary, treatment, removing 40 per cent of BOD; secondary, removing 85 per cent of BOD; and tertiary, removing 95 per cent of BOD. The impact of the discharges from the domestic and industrial sectors, for both the present and estimated future generated waste loads, arising from these levels of treatment was assessed using the mass-balance equation referred to above. The optimum combinations of treatment in the two sectors were assessed by a simple matrix analysis. In the case of lakes where reduction of phosphorus input from point waste sources is considered necessary to control eutrophication, the plans assume that appropriate tertiary treatment will remove between 80 and 90 per cent of the nutrient from the wastes.

Review of plans. The plans proposed that a review process should take place at intervals not exceeding five years. Such review would take into account new data and information and assess the efficacy of measures adopted to control pollution in the preceding period. In order to facilitate this periodic review the plans also incorporated proposals for a continuing programme of investigation and data acquisition. The programmes would be designed to fill gaps in the database and to provide feedback on the impact of measures implemented under the plan.

Limitations of plans drafted by AFF

The plans as presently formulated are mainly concerned with the effects of the BOD in point waste sources (discharges from sewerage systems and from industries). These have been responsible for most of the chronic pollution of a serious nature in the waters covered by the plans and therefore merited priority. The procedure adopted in the draft plans to assess the input of waste discharges is somewhat arbitrary although it is considered to be relatively stringent where BOD inputs, *per se*, are concerned. However, it does not address in any detail the problems resulting from eutrophication in rivers and streams (although these are dealt with in the case of lakes) nor the potential pollution from non-point waste sources, in particular agricultural activities. The main reason for these omissions in the plans was the lack of detailed information at the time of drafting.

With regard to the potentially adverse effects of eutrophication in rivers and streams, AFF commenced work on this in the late 1970s when observations and experiments were carried out on the River Suir (Horkan 1978). The draft plans incorporated a commitment to improving the ability to predict the impact of waste discharges by water quality modelling. This was considered necessary to deal, in particular, with the complexities introduced by eutrophication and its impact on the dissolved oxygen regime. Preliminary studies (McGarrigle 1984) on modelling water quality variations in the River Suir were carried out by AFF in 1984 under a contract with the EC and with support from the local authorities involved. The model has been tested in additional investigations on the Rivers Nore, Slaney and Castlebar. In its present form the model is designed to predict dissolved oxygen levels below outfalls based on the BOD of discharges and measurements of the important physical and biological factors in the receiving waters. Further development, to incorporate a link between nutrient input and plant growth, is necessary to allow a proper assessment of the feasibility of eutrophication control.

The draft plans are further limited, as referred to already, by lack of data or information on various aspects, in particular on waste discharges and river flows. Improvement of the position regarding waste discharges requires an increased level of sampling of effluents, both in the municipal and industrial sectors, as well as effluent flow monitoring. In regard to wastes from non-point sources, particularly agricultural activities, useful information is likely to have been obtained recently by the task forces set up in each local authority area to investigate the pollution-potential on farms. This should allow the control of non-point source pollution to be addressed on a systematic basis in any review of the presently drafted plans or in any new plans to be prepared. In this connection also, there is a need to include information in the plans on any non-agricultural storage of potentially polluting materials, in particular hazardous chemicals used in the manufacturing process or produced as a waste in industrial plants. In the draft plans prepared by AFF this aspect received detailed treatment only in the case of the R. Liffey, where major water abstractions make it a critical consideration. There is a clear need to identify the nature of and the location of all such stores of potentially hazardous materials and to assess the risk which they constitute to adjacent surface and groundwaters and how this risk is or can be minimised.

Deficiencies in flow data must be partly addressed by improvements of the hydrometric network. As pointed out already, many of the gauges installed by the OPW in connection with the planning of drainage works have proved unreliable as a guide to lowflow characteristics. The hydrometric schemes which the local authorities have been putting in place with the collaboration of AFF will remedy these deficiencies. However, many of these schemes are not yet complete. In addition, a considerable period of time must elapse (10-20 years) before useful flow statistics become available for specific gauging stations. This difficulty can be partially circumvented, however, by making flow measurements in periods of drought.

Fisheries are a particularly important consideration in the draft plans. As indicated above, the water quality standards are based on the water quality criteria appropriate to fish because of the general stringency of these criteria. The plans therefore adopt as one of their main objectives the promotion of fisheries since these are an assurance that conditions are acceptable for most other uses. However, the information available on the distribution of the fish stocks in the river systems concerned at the time of drafting of the plans was limited, particularly in respect of the locations of spawning areas. The latter aspect is of critical importance in determining the extent of each river system which should be formally covered by the plan. While, in theory, it seems desirable that all parts of the river system should be covered by the plan, such an approach might, in practice, lead to anomalies and other difficulties. For instance, it is probable that in the very smallest streams the water quality variations, under natural conditions, would deviate from the limits defined in the water quality standards. Furthermore, inclusion of such waters in the plan would also place a heavy burden on the monitoring facilities of the local authorities as sampling would be required for even small streams in remote areas where the risk of pollution is virtually absent. However, it is accepted in the draft plans that any streams where a beneficial use has been identified, e.g. water abstraction or spawning area, should be included in the formal plan area. It may be necessary, as additional information is obtained, on spawning areas in particular, to extend the waters covered by the plan in each case. It should be noted also that all waste discharges, whether within the formal plan area or not, remain subject to the controls laid down in the 1977 Water Pollution Act.

FURTHER DEVELOPMENT OF PLANS FOR IRISH RIVERS

Extension of coverage

In addition to those drafted by An Foras Forbartha other plans have been or are being prepared directly by the local authorities. The latter cover the catchments of the Moy, Corrib, Bandon, Boyne, Feale, Finn, Leannan and Swilly Rivers, the estuaries of the Munster Blackwater and Bandon Rivers and Bantry Bay. A plan for Dublin Bay and the Liffey Estuary, which was commenced by AFF on behalf of the local authorities concerned, will be completed by the Environmental Research Unit. Thus, plans completed or in hand encompass all of the major river systems in the state (the plan for Co. Cavan covers the bulk of the R. Erne system lying in the State). Of the remaining rivers to be dealt with those such as the Lee, Laune and Fergus seem large enough to be the subjects of individual plans. This is hardly justified in the case of smaller rivers which might be more appropriately dealt with as groups of adjacent small catchments.

Of the estuarine and coastal waters not yet considered in draft plans, Cork Harbour is probably the most important; however, the considerable investigative work which has been carried out on the harbour would greatly facilitate the formulation of a plan. Of the other estuarine and coastal waters not so far considered for water quality management planning, areas such as Tralee Bay, Kenmare River, Clew Bay, Sligo Bay, Donegal Bay, Lough Swilly and Carlingford Lough may merit priority as these have been identified by the fishery authorities as having potential for aquaculture.

Other aspects for consideration

In any new plans to be developed or in any revision to be undertaken of existing plans there are a number of aspects which, in line with previous remarks, merit greater attention than was possible at the time, in the late 1970s and early 1980s, when the first plans were drafted by AFF. These aspects are considered below.

Control of eutrophication, particularly in rivers. This will require detailed information on nutrient inputs, particularly phosphorus, from point and non-point sources. The relationship between these inputs and the production of plant biomass, as well as that between the latter and water quality characteristics, particularly dissolved oxygen levels, will need investigation. In this connection the availability of the water quality model developed by AFF may be of assistance. Besides nutrient input reduction, other controls on plant biomass development may need consideration, such as weed clearance or the development of shading vegetation along the river banks.

Control of non-point source pollution. The main potential source of pollution is agricultural activity, particularly livestock manures and silage liquor. Two areas require attention, viz. the interception and storage of waste and the disposal on land. Failure of control in the first area is likely to lead to the entry of high-strength wastes to waters, leading to severe deoxygenation and probably toxic conditions (BOD and ammonia increases). The main effect of lax controls in land disposal of waste is likely to be eutrophication owing to the excess run-off of nutrients. Inclusion of appropriate measures for the control of non-point sources of waste in the plans will require a full inventory of such sources and an assessment of present management arrangements. As mentioned above, the work already put in hand by the local authorities, in cooperation with the fishery boards and local agricultural officers, should provide the necessary information for this aspect of the plans.

The plans should include information on the generation and storage of toxic chemicals and other potentially dangerous substances in the catchment area and specify the measures to be taken to minimise the risk which they represent to surface and groundwaters. In this connection, special consideration should be taken of EC Directive 76/464 (Discharge of Dangerous Substances) and its various daughter directives and Directive 80/86 (Protection of Groundwaters) and the associated national regulations.

Critical flow conditions. A further assessment is required of the critical flow conditions to be used in calculating the waste assimilation capacity for biodegradable wastes in rivers. In the case of the plans drafted by AFF, the 95 percentile flow was used for this purpose. Strong reservations were expressed by the fishery agencies regarding this flow statistic and a more stringent condition, viz. the minimum 7-day sustained low flow with a 15-year return period, was recommended. From an examination of available records, it appears that this flow is of the same order as the long-term minimum flow. It is considered, therefore, that use of this flow would be unnecessarily stringent as far as the assimilation capacity for biodegradable wastes is concerned. However, the examination of the flow records also showed that, in some years, the river flow could be equal to or less than the 95 percentile value for over 50 consecutive days. Thus, at locations where the full assimilation capacity, as presently calculated, is taken up, BOD concentrations would, in theory, vary between 4 and 8 mg/l below the discharge point for long periods in such years. It is not clear if these increased BOD concentrations, *per se*, would have serious effects; the main impact is likely to be an increased growth of bacterial slimes and associated biota on the substratum, with an attendant increased respiratory demand on dissolved oxygen. The water quality model could be used to investigate such situations and to assess the need for adopting a more stringent flow criterion.

Water quality standards. A further assessment of the water quality standards as adopted in the plans drafted by AFF is warranted. This assessment will need to take account of recent national regulations issued in connection with the various EC water directives, e.g. 78/659 (Freshwater Fish) and 76/160 (Bathing Waters). Particular attention is required in the case of the standards for phosphorus in rivers and lakes. The tentative limits specified in the plans drafted by AFF should be reassessed in the light of the considerably expanded database now available. As BOD loads from point sources are gradually diminishing in importance as a cause of pollution in inland waters, eutrophication, particularly phosphorus enrichment, is now the main concern (Toner *et al.* 1986).

Geographical scope. The geographical scope of the draft plans should also be reviewed in the light of any additional information becoming available on beneficial uses, in particular on the extent of known fishery waters. It is possible that this may create a need for seasonally variable water quality standards owing to natural variations in some of the very small streams in which salmonid spawning grounds occur.

Water quality monitoring. Any new or revised plans should incorporate detailed arrangements for adequate water quality monitoring so that a proper assessment can be made of the level of compliance with the standards adopted. There is a general need to evaluate the manner in which available resources are used to monitor water quality and to determine whether the existing sampling regimes need modification so as to optimise the value of the information produced. It is considered that an increased level of sampling frequency at the cost of a reduction in the number of sampling points would be worthwhile.

Finally, it must be emphasised that water quality management plans are not 'once-off' endeavours but should be considered as continuous processes. Periodic review is needed to evaluate the results of policies already adopted and to take account of new information and circumstances. This is particularly necessary in the case of remedial measures undertaken to deal with existing pollution. The efficacy of such measures requires detailed consideration in the review process so that any need for additional controls can be identified. The basis on which to make a reassessment of this and all other aspects of the plan is a system for the continuous collection of data and information.

REFERENCES

ALABASTER, J.S. and LLOYD, R. 1980 *Water Quality Criteria for Freshwater Fish.* London. Butterworths.

COUNCIL OF THE EUROPEAN COMMUNITIES 1975 Council Directive of 16 June 1975 concerning the quality required of surface water intended for the abstraction of drinking water in the Member States. *Official Journal of the European Communities* L194, 26-31.

COUNCIL OF THE EUROPEAN COMMUNITIES 1976 Council Directive of 8 December 1975 concerning the quality of bathing water. *Official Journal of the European Communities* L31, 1-7.

COUNCIL OF THE EUROPEAN COMMUNITIES 1978 Council Directive of 18 July 1978 on the quality of fresh waters needing protection or improvement in order to support fish life. *Official Journal of the European Communities* L222, 1-10.

DOWNING, D. and SESSIONS, S. 1985 Innovative water quality-based permitting: a policy perspective. *Journal of the Water Pollution Control Federation* 57, 358-65.

HORKAN, J.P.K. 1978 *Interim Report on Eutrophication and Related Studies of the Thurles Area of the River Suir (May-July 1978)*. Dublin. An Foras Forbartha, WR/R11.

LEE, G.F., JONES, R.A. and BROOKS, W.N. 1982 Water quality standards and water quality. *Journal of the Water Pollution Control Federation* 54, 1131-8.

McGARRIGLE, M.L. 1984 *Development of a Water Quality Model for Irish Rivers*. Dublin. An Foras Forbartha, WR/R16.

OECD (Organisation for Economic Co-operation and Development) 1982 *Eutrophication of Waters. Monitoring, Assessment and Control*. Paris. OECD.

TONER, P.F. 1987 Freshwater pollution in Ireland - does it present a problem? In M. Murphy (ed.), *Lake, River and Coastal Pollution - can it be contained?*, 15-26. Cork. Sherkin Island Marine Station.

TONER, P.F., CLABBY, K.J., BOWMAN, J.J. and McGARRIGLE, M.L. 1986 *Water Quality in Ireland. The Current Position. Part One: General Assessment*. Dublin. An Foras Forbartha, WR/G15.

TRAIN, R.E. 1979 *Quality Criteria for Water*. London. Castle House Publications.

WARN, A.E. and BREW, J.S. 1980 Mass balance. *Water Research* 14, 1427-34.

WATER POLLUTION ADVISORY COUNCIL 1983 *A Review of Water Pollution in Ireland. A Report to the Council by An Foras Forbartha*. Dublin. Water Pollution Advisory Council.

In: Steer, M.W. (ed.) 1991 *Irish Rivers* : *Biology and Management* pp 255-266.
Royal Irish Academy, Dublin.

THE FISHERIES RESOURCE : MANAGEMENT CONTROLS IN ENGLAND AND WALES

Richard C. Cresswell*

National Rivers Authority, Welsh Region, Cardiff

ABSTRACT

The fisheries resource in England and Wales is subject to both legal and illegal exploitation, and to a range of environmental pressures including pollution, abstraction and river engineering. Management controls are essential if the resource is to be preserved.

The administrative and legislative frameworks which allow for the management of fisheries in England and Wales are briefly outlined. Ways by which some of this legislation is utilised to achieve management control over exploitation and environmental factors affecting the fisheries resource are described. Examples are drawn from Wales and primarily concern measures to control the exploitation of salmon and sea trout stocks and measures to prevent environmental degradation. The latter, whilst further protecting the migratory salmonids, are more widely applicable and are generally beneficial to the fisheries resource as a whole.

AREAS OF RESPONSIBILITY

The responsibility for the overall management and regulation of salmon, trout, freshwater and eel fisheries in England and Wales rests with the National Rivers Authority (NRA), which came into being as a result of the Water Act 1989. The Authority is divided into ten regions, each based on the area of a former regional water authority (Fig. 1).

----

* The views expressed in this paper are those of the author.

The NRA's powers to enforce legislation relating to salmon and sea trout in coastal waters extend out to the 6-mile limit. The regulation of sea fisheries is generally the responsibility of independent sea fisheries committees whose powers extend out only 3 miles. The NRA does, however, have the powers of a sea fisheries committee in some significant areas, e.g. Welsh region - Dee Estuary and a large part of the Severn Estuary.

Fig. 1. Regions of the National Rivers Authority.

LEGISLATION

The statutory duties of the NRA with respect to fisheries are largely defined by the Salmon and Freshwater Fisheries Act 1975 (1975 Act), and have been reaffirmed by the Water Act 1989. The 1975 Act is a consolidation of legislation going back to the 1923 Act, which in turn consolidated legislation going back to the 1860s. Not surprisingly there are deficiencies in the 1975 Act, the basis of which was conceived over a century ago when the pressures on the resource and management requirements were different. Nevertheless, the 1975 Act and associated bye-laws still provide most of the basic controls on methods of illegal fishing, close seasons, close times, size limits and also requirements for fish passes, fish introductions etc.

The 1975 Act has been amended and supplemented by various pieces of legislation, most notably the Salmon Act 1986 which contained provisions designed to control the trade in illegally caught fish and powers to introduce bye-laws to overcome the dichotomous legislation relating to the capture of migratory salmonids and marine fish in coastal waters.

In addition to primary fisheries legislation there are a number of other areas of law which have been developed under separate statutory provisions but which allow important management controls to be exercised (Table 1). This legislation may appear wide-ranging, but its effects are fundamental to modern fisheries management.

One of the most important provisions of the 1975 Act (Section 28(1)(a)), confirmed by the Water Act, 1989 (Section 141), is the duty which it imposes upon the NRA:

> "to maintain, improve and develop salmon fisheries, trout fisheries, freshwater fisheries and eel fisheries".

This rather open-ended responsibility, although necessarily restricted by available resources, enables the NRA to utilise a range of legislation to benefit fisheries where this is compatible with the main purpose of that legislation.

Table 1. Legislation of importance to fisheries management.

Salmon & Freshwater Fisheries Act, 1975	
Diseases of Fish Acts, 1937, 1983	basic
Import of Live Fish (England & Wales) Act, 1980	fisheries
Salmon Act, 1986	regulations
Theft Act, 1968	
Water Resources Act, 1963	powers of
Salmon & Freshwater Fisheries Act, 1975	water
Police & Criminal Evidence Act, 1984	bailiffs
Justices of the Peace Act, 1361	binding over of offenders
Water Resources Act, 1963	water abstraction
Water Acts, 1973, 1989	powers & duties of NRA
Control of Pollution Act, 1974	control of pollution

MANAGEMENT INFORMATION

During the past century, various positive attempts have been made throughout England and Wales to manage and improve fisheries. However, attempts were often misdirected and it was not until the creation of the River Authorities in 1963 that management agencies made much effort to monitor the effects of their actions. Since then there has been an increasing recognition that effective management demands knowledge of the status of the resource. An awareness of whether stocks are stable, decreasing or increasing is essential for defining management needs, setting priorities and allocating resources.

Resource assessment

Historically, the levels of catch returns have been cited as proof of changes in fishery status, and in the absence of more accurate scientific data these records currently retain their importance. From 1976 the Welsh Water Authority (WWA) sought to enhance the quality of returns from anglers and commercial fishermen through improved administration, recording and reporting procedures and by the issue of postal reminders to non-respondents. The return rate from commercial fishermen is now almost 100%, and from anglers has risen from 10-15% to a fairly consistent 50-60%. The consistency and increased completeness and general accuracy of the catch data make them a more reliable indicator of fishery status. Whilst some data on fishing effort are already being collected by voluntary returns, a provision of the Water Act 1989 enables the NRA to promote a bye-law to require fishermen to include such information on their catch returns. A review of existing information is being undertaken to ensure that the right sort of effort data are collected.

Figure 2 provides an example of annual rod catch data for the Welsh region, and shows a dramatic 50% increase in sea trout catches in 1987. One major benefit of catch data is to focus attention on areas where such changes may be occurring.

Awareness of the risks of using catch data as an index of stock abundance led to the establishment of a range of monitoring programmes. A 5-year programme of redd counting on all significant fisheries yielded valuable information on the distribution and relative importance of spawning areas and the significance of obstructions to the utilisation of potential nursery streams. It also established that redd counts, besides being costly and labour-intensive, could not be used as a reliable indicator of stock abundance, particularly on rivers containing both salmon and large sea trout.

In response to a paucity of baseline data on juvenile salmonid stocks, a monitoring programme was initiated in 1985. The main effort has been targeted at those catchments supporting major commercial and recreational fisheries and involves annual sampling of key sites. Baseline information on other catchments is obtained less frequently under a rolling programme and

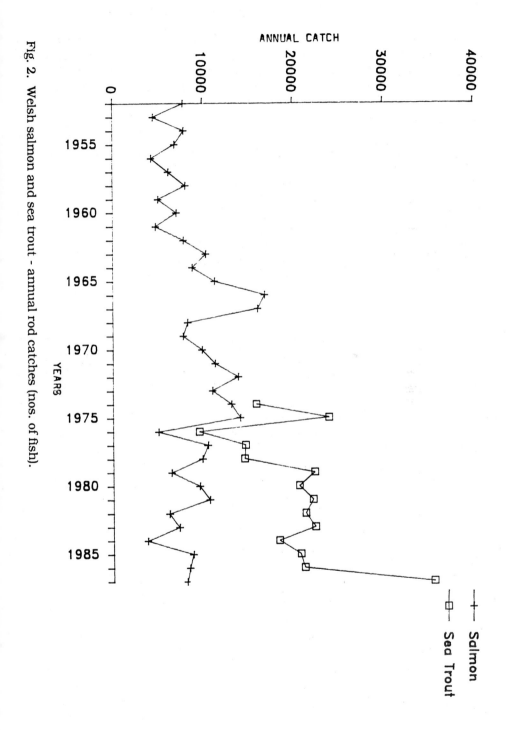

Fig. 2. Welsh salmon and sea trout - annual rod catches (nos. of fish).

will eventually be available for all (more than 50) catchments in the region. Over the period 1985-7, an average of 109 quantitative surveys and 257 semi-quantitative surveys have been carried out per year. In order to summarise raw data on population densities, a simple classification system has been established (WWA 1988a). This allows areas of poor population to be readily highlighted and for appropriate management action to be targeted to those areas. This system operates throughout Wales and it is proposed to extend its use into England.

Fish counters have produced few useful data on Welsh rivers, primarily because of operational unreliability and doubts about their accuracy and value on rivers containing salmon and large sea trout. Although some progress has been made in the last couple of years with the latest designs, constraints on siting of counters will remain. The value of fixed traps is acknowledged, although the associated capital and revenue costs are comparatively high. One such facility is currently being constructed on the lower Welsh Dee, and another has been constructed on the River Taff as part of the Cardiff Bay Barrage Fisheries monitoring programme.

MANAGEMENT CONTROLS

Exploitation

a) *Licensed fishery.* The legal exploitation of fish in England and Wales is regulated by a system of licensing operated by the NRA. The number of fishing licences issued is unrestricted unless a limitation order is in force. There are no such orders on rod and line fishing, but the number of netting licences is limited in most regions of England and Wales.

In the Welsh region, rod licence sales for salmon and sea trout have in recent years fluctuated around 16,000 per annum (1987 - 17,632) compared with a relatively stable number of commercial net licences (approximately 175). The comparatively small number of net licences accounts for approximately 50% of the reported catch of migratory fish, but brings in only a modest income (Table 2). Enforcement effort on the licensed fishery, although minimal compared with that on the unlicensed fishery, is most effectively directed at the commercial nets, which licence for licence represent a much greater proportion of the declared catch.

Table 2. Salmon and sea trout: average annual rod and net licence sales and catch return, 1983-5.

| | Licences | | Catch return | |
	No. sold	Value	Salmon	Sea trout
Rods	15,748	243,594	6,683	20,939
Nets	173	11,988	4,072	8,199

Legislation allows some degree of selectivity, within strict guidelines, in the allocation of net licences, but provides no control over the appointment of endorsees who help to work the net. This was a major problem until the Salmon Act 1986 made it a requirement that the licensee be present whenever his net is fished. Previously, convicted poachers, appointed as endorsees by a licence holder, could work a net unsupervised.

For many years it was a matter of great concern that, in the absence of alternative candidates, a commercial net licence could not be withheld from a convicted poacher unless he had been convicted twice and had, as a result, been disqualified from holding a licence by the courts. The maximum disqualification period was 1 year. As a result of the Water Act 1989, a disqualification for up to five years can now be imposed upon a first conviction.

b) *Illegal fishery.* Illegal fishing is a serious problem and in certain areas the take is considered to equal that of the licensed fishery. In recent years there has been a decline in winter poaching (use of lamp and gaff on spawning streams) in favour of netting activity during summer months. On inland waters such netting is practised by mobile gangs, whose members are often involved in other criminal activity. This type of illegal fishing is being countered by an increasingly mobile bailiff force, backed up with trained dogs and modern surveillance equipment.

Illegal netting of salmon and sea trout is also a major problem in coastal and estuarine waters where the activity is conducted under the guise of fishing for sea fish, and enforcement is hindered by dichotomous legislation relating to the capture of salmonids and sea fish in coastal waters.

c) *Fixed nets and drift nets.* Fixed nets set on beaches take a large number of salmonids around the coast of north and west Wales, whilst illegal drift netting is a serious problem in the Dee and Severn Estuaries. The seriousness of the fixed net problem is illustrated by a survey carried out from 18 April-31 July 1984 on a beach in north Wales. A total of 75 out of 212 tides were checked and an average of 7 nets were found to be operating each tide. The total catch of these nets was 54 salmon, 237 sea trout and 40 sea fish (other than dogfish). In such circumstances it would be unreasonable for a fisherman to say that his intended quarry was sea fish. During the same period the 6 licensed commercial salmon nets on the Afon Conwy caught 165 salmon and 188 sea trout, and the licensed rods only 26 salmon and 30 sea trout.

The use of fixed nets is controlled by Section 6 of the Salmon and Freshwater Fisheries Act 1975 which may have been inadvertently altered during the consolidation of the 1923 Act. Prior to 1975 the placing or use of a fixed engine (most often a net) *for the purpose of catching salmon* etc. was an offence. The 1975 Act divided the former crime into two sections:

a) placing a fixed engine in any inland or tidal water;
b) using a fixed engine for taking salmon etc.

A case in 1984 (Champion v. Maughan and Groves) tested the law, and it was held that placing a fixed engine had become an absolute offence. This effectively made all fixed nets illegal, including those set by genuine sea fishermen. It is unlikely that this was the intention of parliament, so the former water authorities, and latterly the NRA, have adopted an even-handed approach, devising informal agreements with local sea fisheries committees; areas and times when fixed nets could be used were defined.

The Salmon Act 1986 sought to clarify and legalise the situation. Section 6 of the 1975 Act was further amended so that, "placing or using a fixed net in any inland or tidal water" became an absolute offence. It did, however, also make provision for bye-laws to be formulated which would authorise the placing of fixed nets at certain times and locations (principally when and where salmonids would not be taken in any great numbers). The Salmon Act 1986 also made provision for bye-laws to be formulated under sea fisheries legislation, specifically for the protection of migratory salmonids. These new bye-law-making provisions are an attempt to overcome the problems caused by the dichotomous legislation.

The formulation and introduction of bye-laws as described above, whether by the NRA or the sea fisheries committee, requires, in law, the consent of the other party. In practical terms, the NRA has to persuade the sea fisheries committees to introduce bye-laws (which may affect sea fishing) to protect salmonids and in turn the sea fisheries committees have to gain consent from the NRA for bye-laws to permit fixed nets which may take salmonids; the legislation has therefore created a situation where lengthy negotiation has been necessary to obtain the desired controls. Whilst this consenting arrangement ensures consideration of both parties' interests, it is administratively time consuming and is in addition to the normal public consultations.

The number of salmonids taken in fixed nets around the Welsh coast has been dramatically reduced since the test case in 1984 and the introduction of the informal agreements. However, in many areas of England and Wales, the bye-laws, permitted by the Salmon Act 1986, which would legalise the fixed net situation and control other forms of netting to protect salmon are still subject to discussions between NRA and sea fisheries committees.

Meanwhile, as legislation proceeds to control the use of fixed nets, the "illegal fishermen" have adopted forms of beach netting which continue to stretch the definition of a net which is fixed and stationary! For this reason, the use of drift nets close to the shore will also require control by bye-law. In those areas, such as the Dee and Severn Estuary, where the water authority is the sea fisheries committee, such bye-laws are being promoted.

Fish dealers

Illegal fishing, both in the rivers and the sea, is being contained but not beaten and this situation will remain so long as trade in illegally caught fish continues relatively unimpeded. The Salmon Act 1986 introduced two provisions intended to control this trade in illegally caught fish. The first would have enabled the introduction of a system of dealer licensing, intended to strengthen control on the illegal trade. However, such a system has not been promoted; after an extensive consultation period, the government concluded that, in the light of the administrative and enforcement burdens involved, such a scheme would not prove workable or beneficial.

The second was to create the offence of "handling in suspicious circumstances", effectively requiring anyone buying salmon to establish that it had been taken legally. The basic offence of handling fish in suspicius circumstances is set out as follows: "a person shall be guilty of an offence if, at a time when he believes or it would be reasonable for him to suspect that a relevant offence has at any time been committed in relation to any salmon, he receives the salmon, or undertakes or assists in its retention, removal or disposal by or for the benefit of another person, or if he arranges to do so."

Having proved that the person was in possession of or handled the salmon in some way, the prosecution must show that the circumstances were such that *a reasonable man* would have had reason to *suspect* that the fish had been taken illegally. If this is successfully proved, then the defendant would be able to acquit himself only by *proving* that the salmon *had not in fact* been taken illegally.

The tendency to deal in frozen salmon and the inability to identify individual fish restricts the success of this piece of legislation in controlling the "buyers". However, it has proved a useful charge against persons caught in possession of fish under suspicious circumstances (e.g. on the river bank at night) but where for evidential reasons a charge of illegal fishing could not be pressed.

Efforts have been made to enhance the chance of successful prosecution against fish dealers or hotels who purchase illegally caught fish. Information leaflets, advising dealers to obtain proof that a fish has been caught legally, have been distributed often by recorded delivery. It is hoped that this will not only dissuade many from buying suspect fish but will also reduce the effectiveness of a defence claiming that the buyer had not thought to ask about the origin of the fish.

Despite the new legislation, control of the illegal trade in salmon remains a major problem. Many fisheries officers, responsible for the enforcement of fisheries legislation in England and Wales, maintain that the individual tagging of salmon and sea trout carcasses is essential so that individual fish can be identified and trade in illegally caught salmon be effectively controlled. Such a tagging scheme does not, however, feature in the current programme of proposed legislation.

Environment

Although fisheries protection is often regarded purely in terms of the control of legal and illegal exploitation, control of factors affecting the aquatic environment are also of fundamental importance in safeguarding the future of fish stocks.

Three major factors over which controls are exercised are water quality, abstraction and river engineering.

(a) *Water quality*. Water quality throughout Wales is generally good with 70% of river and canal lengths (total - 4,802 km) achieving or exceeding long-term objectives in 1986/87 (WWA 1987). The extent to which problems exist varies as does the relative importance of different pollutants, e.g. coal washings in south Wales, agricultural problems in south-west Wales and metals in mid and north Wales. The acidification of surface waters in the upland areas of mid and north Wales is of major concern and subject to extensive investigation (Milner and Varallo 1990).

The consequence of pollution may be intermittent fish kills or persistent poor water quality which prevents fish from inhabiting or spawning in an area. The latter problem may be identified by the fisheries monitoring programme, but the implementation of normal water quality controls should prevent it occurring. The major problem is intermittent pollution which is more difficult to detect and remedy.

If a pollution incident occurs involving a fish mortality, it is the responsibility of the NRA to ensure that whenever possible the fishery receives adequate compensation and also to oversee any reinstatement work. Increasingly, riparian owners have left the NRA to claim compensation on their behalf and to restore the fishery, whilst they claim for the loss of the amenity in the intervening period. In a recent case, compensation monies were utilised to improve hatchery facilities on an affected river system, the existing facilities themselves being the result of a £100,000 compensation payment in respect of an earlier fish kill caused by a mine discharge.

(b) *Abstraction*. Abstraction from rivers must be effectively controlled both to protect fish populations and the quality of fishing. A working group set up within the WWA to examine the management of water abstraction and its effects on the environment produced guidelines recommending that the minimum environmental flow for a good-quality river be its 95 percentile flow (i.e. flow exceeded 95% of the time on average) and also that a limit be put on the artificial duration of this low flow (WWA 1988b).

Close liaison between fisheries staff and those drawing up control curves and rules for compensation releases from water supply reservoirs can lead to mutual benefits. Devising the operating rules of reservoirs so that compensation releases are minimised when they are of least value to the environment has the benefit of increasing the security of water supply. This in turn reduces the risk of drought orders being sought and the associated detrimental effects on rivers and their fisheries. Cooperation in the conservation of water supplies in this way can enable certain volumes of stored water to be set aside to be used for fisheries management purposes (Lambert 1988).

The success of such an appoach was highlighted on one river/reservoir system on the Lleyn Peninsula, north Wales, during 1988. Reservoir storage had been carefully managed so that, despite dry weather, sufficient volume of water was available to supplement the compensation flow into the river, which had fallen to a critically low level for the time of the year. This supplemented compensation flow kept the river fresh for several weeks before rainfall occurred.

(c) *River engineering.* In the past, river engineering schemes have destroyed areas of riverine habitat affecting all stages of salmonid development. However, liaison between fisheries staff and engineers, together with a responsibility to further conservation in the discharge of authority functions, has led to a more sympathetic treatment of sensitive areas. River engineering works carried out by agencies other than the NRA require land drainage consent. Works covered by these consents often have implications for fisheries, and relevant staff are able to have an input to ensure that proposals are not detrimental.

Major initiatives which are likely to have a significant but undetermined effect on fisheries, such as the siting of a power station or creation of a tidal barrage, are necessarily subject to detailed impact assessment schemes. Properly conceived pre- and post-scheme impact studies are required to:

i) assess the likely effects on fish, fisheries and fishing;

ii) indicate how predicted losses may be minimised or otherwise mitigated;

iii) determine whether or not the construction and operation of the scheme successfully avoided damage.

Impact assessment schemes and resultant mitigation are generally conceived and implemented on the principle of "promoter pays". This concept applies whether the scheme is an internal land drainage operation or a major construction project. The principle is illustrated by a scheme to construct a road tunnel by dredging across the Afon Conwy estuary in north Wales. The Welsh Office, promoting the scheme, are not only financing the impact assessment but also the construction of a fish pass on Conwy Falls, a major obstruction. The fish pass will open up a significant part of the catchment to migratory salmonids and is financed in lieu of compensation for any possible loss of fishing amenity which might have been caused. If the fish or fishery is proved to be detrimentally affected, further compensation for restocking etc. will be payable.

Negotiated agreements of this kind are becoming more common as public awareness of the need to protect the environment has grown.

CONCLUSIONS

Most management controls on fisheries in England and Wales have their roots in legislation, but it has been the associated liaison and consultation procedures which have enabled maximum benefits to be gained in recent years, particularly in respect of environmental protection. There is no room for complacency, however.

-Illegal fishing continues to be a problem and one which could still be helped to a significant extent by updating and strengthening legislation.

-The generation of reliable scientific data remains fundamental to the efficient and effective management of the resource.

-Finally, as people attain more leisure time, there is an increasing demand for recreational facilities. Until relatively recently, our rivers were enjoyed almost exclusively by anglers and, of course, by the native wildlife. Now a great variety of activities, including canoeing, rambling, rowing and rafting is beginning to make increasing claims on the riverine resource. The future coordination of these activities and the avoidance of conflict between participants present a new challenge in terms of management control. Legislation will be required to form a basis for management, but liaison and consultation with an informed public will allow the greatest overall benefits to be achieved.

REFERENCES

LAMBERT, A. 1988 An introduction to operational control rules using the 10-component method. *British Hydrological Society*, Occasional Paper No. 1.

MILNER, N.J. and VARALLO, P.V. 1990 Effects of acidification on fish and fisheries in Wales. In R.W. Edwards, A.S. Gee and J.H. Stoner (eds), *Acid Waters in Wales*, 121-43. Dordrecht. Kluwer.

WELSH WATER AUTHORITY 1987 *Report and Accounts 1986/87*. Brecon. Welsh Water Authority.

WELSH WATER AUTHORITY 1988a *Regional juvenile salmonid monitoring Programme*. Annual Report 1987. Brecon. Welsh Water Authority.

WELSH WATER AUTHORITY 1988b *Guidelines for management of water abstraction in the environment*. Brecon. Welsh Water Authority.

In: Steer, M.W. (ed.) 1991 *Irish Rivers : Biology and Management*, pp 267-277. Royal Irish Academy, Dublin.

LEGAL CONTROLS OF POLLUTION

Sean McMorrow

Barrister at Law

INTRODUCTION

The aim of this paper is to outline the main legal measures that are available to control pollution in rivers in this country. It is intended as a guide for biologists who may become involved in legal proceedings, particularly as expert witnesses or in their capacity as scientific advisors in relation to the terms of licences issued as part of the planning or pollution control systems.

The law dealing with pollution of rivers is mainly derived from three sources: the *common law* (i.e. cases and judgements of the courts), *statute* and the *E.E.C.* Generally the common law is concerned with riparian owners seeking to protect and preserve their proprietary rights against pollution through the intervention of the civil courts by way of damages and injunction. The statute law endeavours to prevent pollution by prescribing safeguards and causing offenders who pollute to be dealt with before the criminal courts.

The common law approach to pollution is founded on the right of the owner of land on the banks of the stream to receive the flow of water past his land in its natural state of purity. Anyone fouling the water infringes the right of property of the riparian owner, who can maintain an action for nuisance against the infringer, unless that person has acquired a right to pollute the water.

Rights to water are incidental to ownership of the adjoining land. While the riparian owner through whose land a stream or river flows may be the owner of the banks and bed of the watercourse, he does not own the water. Water flowing from its source to the sea is public property. The riparian owner is entitled to use it for his domestic and agricultural needs; industrial companies can use it for their manufacturing purposes; but while they have the right to use the water flowing past their property, they must not sensibly, that is, appreciably, diminish the *volume* or *quality* which flows downstream to other riparian owners. Since flowing water is public property, the right of the riparian owner, and of anglers and other users in general, is essentially one of access, so that where a riparian owner owns one or both banks of a river, he can lawfully prevent anyone from having access to the water over his land. It should be noted here that the fish that live and swim in flowing water are wild creatures and are the property of nobody until they have been caught and are capable of being carried away. This is so even where an angling club has spent a great deal of money in the purchase of fish to re-stock the river. From the moment the fish are released in the river, they become wild creatures. If they are put in private ponds from which they can-not escape, they belong to the owner of the ponds.

Pollution at common law means the addition of anything to water which affects its natural qualities and thereby results in the riparian owner not receiving the natural waters of the stream.

Who may sue?

Where an injury has been done to the private rights of a person, whether tenant or landlord, he is entitled to damages, and where in such a case a serious injury is apprehended, an injunction may be granted to restrain the person in default. A licensee of the riparian owner using the water may maintain an action for fouling the water. A person who has no proprietary title to the water and no leave or licence to use it cannot maintain a civil action for pollution. This applies to non-riparian owners who have no rights with respect to the water.

When a civil action lies

Pollution is simply another aspect of the ancient laws of trespass and nuisance. It is an infringement of the owner's right to enjoy the use of his property without unlawful interference. A civil action for pollution can be maintained without the plaintiff proving that he has sustained actual dam-age on the basis of interference with his rights. He is entitled to have the water in a pure condition and has a right to take the persons causing the pollution, one by one, and prevent each from discharging his contribution to that which becomes, in the aggregate, a nuisance.

E.E.C. CONTROLS

Thanks to our entry to the E.E.C. in 1973, statute law on control of pol-lution has developed significantly over the past decade or so. While the Treaty of Rome does not deal expressly with the environment, the Council of Ministers of the Community have emphasised the importance of an effective campaign to combat pollution and nuisances and the protection of the envi-ronment as an essential part of promoting a harmonious development of eco-nomic activities. Since 1973, the Council has approved three programmes of action on the environment - 1973 - 86. Most recent Irish measures for pollu-tion control have resulted from attempts to comply with the first and second action programmes. These E.E.C. programmes have been implemented by regulations, directives, decisions and recommendations.

Regulations are applicable here as soon as they are promulgated by the

Community authorities and do not normally have to be incorporated into Irish law. *Directives* are binding only as to the "objective to be achieved" and usually have to be incorporated into our law. *Decisions* are binding on those to whom they are addressed. *Recommendations* are not binding, though they may be morally and politically persuasive.

Directives have been incorporated into our law by statute, by delegated legislation, and by administrative action, e.g. Statutory Insts. under E.C. Acts 1972-3: circular letters sent by Ministers to various pollution control authorities on methods of implementation. These last are considered unsatisfactory. The obligations are not known to the public and the letters are confidential between sender and recipient.

Water quality and emission standards

The only mandatory water quality and emission standards are those prescribed in EEC directives applicable in Ireland. The Minister for the Environment has powers to prescribe standards under Section 26 of the Local Government (Water Pollution) Act 1977 but he has never used these powers. A Technical Committee on Effluent Water Quality Standards established by the Minister has issued Memorandum No. 1 on Water Quality Guidelines, and recommended standards in this document are frequently applied by pollution control authorities.

The following EEC directives have been implemented in Ireland by circular letters:

(i) Directive 75/464 on pollution caused by certain dangerous substances discharged to the aquatic environment of the Community;

(ii) Directive 75/440 concerning the quality of surface water intended for the abstraction of drinking water in the member states;

(iii) Directive 80/68 on the protection of groundwater from certain dangerous substances;

(iv) Directive 80/778 relating to the quality of water intended for human consumption;

(v) Directive 79/869 concerning methods of measurement and frequencies of sampling and analysis of surface water intended for the abstraction of drinking water in the member states;

(vi) Directive 80/777 on the approximation of the laws of the member states relating to the exploitation and marketing of mineral water;

(vii) Directive 78/659 on the quality of fish waters needing protection or improvement in order to support fish life;

(viii) Directive 76/160 concerning the quality of bathing water;

(ix) Directive 85/337 concerning Environmental Impact Studies.

STATUTORY CONTROLS

The most important legislation on water pollution is the Local Government (Water Pollution) Act 1977. A number of other statutes are also relevant to water pollution control. These include the Public Health (Ireland) Acts 1878-90, the Fisheries Acts 1959-80, and the Local Government (Planning and Development) Acts 1963-83.

Public Health Acts

The Public Health Act 1878 creates a number of offences in relation to water pollution - the most significant is the statutory nuisance created by Section 107 whenever any pool, watercourse, etc., is so foul as to be a nuisance or injurious to health. The Act places a duty on sanitary authorities to inspect their district for such nuisances and enables abatement and prosecution. Any person aggrieved may prosecute - Section 19 of the 1878 Act. The sanitary authority may not discharge sewage or filthy water into any natural stream or watercourse, etc., unless it is freed from "all excrementatious or foul or noxious matter".

Section 15 vested sewers in sanitary authorities who are obliged to permit any individual to empty his drains into them on giving notice to and complying with directions of the sanitary authority as to manner of connection - once in the public sewer, the sewage becomes the property of the sanitary authority, which is then responsible for treating and disposing of it. Offensive trades are controlled by the Act (Section 128).

Fisheries Acts 1959/80

In practice, the majority of prosecutions for water pollution are brought under Section 171 which provides that any person who:

(a) steeps in any waters any flax or hemp, or

(b) throws, empties, permits or causes to fall into any waters any deleterious matter, shall unless such act is done under and in accordance with a licence granted by the Minister (for Fisheries) be guilty of an offence. Maximum fine is £500 and/or six months' imprisonment. "Deleterious matter" is defined in Section 2 of the 1962 Act - any substance which *is liable to* render waters poisonous or injurious to fish or fish-life/food or spawning of fish. Licences are held by some sanitary authorities, food processing and chemical firms - under 30 in number.

Section 172 (as amended by the 1962 Act) makes it an offence to discharge deleterious matter contained or conveyed in a receptacle which is within 30 yards of any waters - unless done under licence. No licence has ever been issued. Fine: £500/+6 months + £50 per day/maximum £600.

This is intended to control the washing out of spray and slurry tankers/ creamery cans. Bye-laws under Section 9 of the 1959 Act prohibit the discharge of effluent from sand or gravel washing plants into any waters without the consent of the Minister for Fisheries.

Local Government (Water Pollution) Act 1977*

Local authorities have primary but not exclusive responsibility for ensuring the preservation, protection and improvement of water quality. The Act applies to all waters within national jurisdiction, including the sea and aquifers.

Pollution is defined (Section 1) in broad terms of "polluting matter" and its potential to injure various uses of water - for fish life, public health, domestic, industrial, agricultural or recreational.

Section 3 provides that a person shall not *cause or permit* any polluting matter to enter waters. This prohibition does not apply to trade or sewage effluents which are subject to control under other sections as well as other specified exceptions.

Section 4 prohibits the discharge of any trade or sewage effluent *to any waters* except in accordance with a licence. "*Trade effluent*" means an effluent discharged from premises used for carrying on any trade or industry, including mining, but not domestic sewage or storm water. "Trade" includes agriculture, aquaculture, horticulture and any scientific research or experiment. Sewage effluent is treated or untreated sewage. The section applies only to discharges of trade or sewage effluents which enter waters from "*any works, apparatus plant or drainage pipe used for their disposal to waters*". The Section is directed *at point sources only*. Run-off of effluents, slurry, etc., from lands are not controlled by Section 4 but by Section 3.

Section 16 controls by licence the discharge of trade effluents *to sewers*.

Section 27 provides for control of pollution by water-borne craft by means of regulations. The licensing process under Sections 4 and 16 is similar to the planning permission process whereby public participation and appeals to An Bord Pleanala are provided for. Accordingly, the role of the biologist can be of major importance with regard to the operation of this system in advising on precautions to be taken in the issue and review of licences. Consultation with fisheries and other authorities on effluent licence applications enables such interests to be protected.

Enforcement. Breaches of Sections 3 or 4 involve penalties on summary conviction (in district court) not exceeding £250 and/or 6 months' imprisonment and £5000 on indictment (in circuit court). In addition, under Section 10, an order of the district court may be obtained directing the polluter in breach of Sections 3 or 4 to mitigate or remedy any effects of such breach within a specified time.

On failure to comply with the order or notice, the local authority may act to stop the contravention or remedy its effects. The cost of remedying the

----

*See Appendix 1 for outline of proposed amendments in Pollution Bill 1989.

272

damage done may run into several thousands of pounds for restocking fisheries, etc. Here again, the biologist will play a major role.

Section 11 provides that a local *authority* or *any person* may apply to the
high court to prohibit the continuance of a breach of Sections 3 or 4.

Section 12 enables a local authority to serve notice on a person having
custody or control of polluting matter on his premises to take specified measures to prevent this matter from entering waters. Failure to comply is subject to a fine of £250 and the authority may take any necessary steps to deal
with the problem and recoup the cost from the person.

Section 13 deals with cases of emergency and enables the local authority
to take urgent measures and recover the costs, as in Section 12.

Section 14 obliges a person to notify the local authority of an accidental
spillage, subject to penalty of £250.

Section 15 enables local authorities to make water quality management
plans for any waters in its area. Such plans must contain objectives for the
prevention and abatement of water pollution. It also provides for public participation by means of representations. Applications for licences to discharge
effluent to waters are considered in the light of these plans.

Under Section 23 the local authority may seek information regarding
abstractions and discharges.

Section 26 empowers the Minister for the Environment to make regulations prescribing:

    (a) quality standards for water and sewage effluents;

    (b) standards for methods of treatment of such effluents.

No such regulations have yet been made. Therefore, the only mandatory
water quality standards here are those prescribed in EEC directives. A Technical Committee on Effluent and Water Quality Standards, appointed to
advise the Minister on water quality and emission standards, has issued
guidelines to assist local authorities in dealing with development proposals
which may affect water quality. They are confined to recommending quality
objectives for the protection of fisheries and of man, insofar as he consumes
water or the fishery resources therein.

Quality standards contained in Directive 78/659/EEC on the quality of
freshwater needing protection or improvement, in order to support fish life,
have been adopted as national standards, pending the provision of standards
by the Minister.

Section 27 provides for regulating disposal of sewage from water-borne
craft by local authorities.

Local Government (Planning and Development) Acts 1963-83

While not designed for pollution control as such, the Planning Acts have
played a significant part in preventing water pollution through the regulation
and control of land use and development. This has occurred mainly in areas
of industrial and urban expansion.

Welcome features of the planning code are the recognition of the importance of the environment and the facility for public participation in the system. This in turn enhances the position of the professional expert in the environmental field. The biologist may be asked to advise on the likely effects of a proposed use or development of land or buildings, the subject of an application for planning permission, either on behalf of the applicant or of an objector. His advice may therefore influence the decision made. His experience and skill may be sought in preparation of area development plans and in respect of special amenity orders. Therefore, a brief outline of some relevant provisions of the Planning Acts may be of assistance.

Planning Act 1963

Section 24:   planning permission is required for any *"development"* of land which is not exempted (as defined in Section 4 as amended).

Section 3:   defines "development" as the carrying out of *works*, on, in or under land, or the making of any material change in the *use* of any structures or other land.

Section 2:   defines "works" as including any act, or operation of construction, excavation, demolition, repair or renewal.

Sections 26/27: provide that when examining a planning application, the planning authority is obliged to consider the "proper planning and development of the area, *including the preservation and improvement of the amenities thereof,* regard being had to the provisions of the development plan or of any special Amenity Order relating to the said area."

Under 1983 regulations, where pollution of waters may occur, the application is to be referred to the Fishery Board, as well as other interested bodies, as specified. Permission may be refused if the pollution implications are incompatible with the proper planning and development of the area.

Requires every local planning authority to make a plan indicating development objectives for its area. Mandatory objectives for rural areas include development and renewal of obsolete areas, improving and extending amenities.

The Development Plan operates as the framework within which applications for planning permission are examined and decided.

Sections 42/43: empower a local authority to make a Special Amenity Order for an area of outstanding national beauty or of special

T

recreational value or a need for nature conservation.

Section 46:   makes provision for Conservation Orders to preserve from extinction or otherwise protect any fauna or flora in an area.

Section 25:   (as amended by Section 39 of the 1976 Act and the 1977 Regulations) provides for Environmental Impact Studies where planning permission is sought for development costing more than £5m (EC Directive 85/337 - 3/7/88).

Section 27:   of the 1976 Planning Act empowers *any person* to seek a high court order prohibiting an unauthorised development causing water pollution or enforcing pollution control conditions attached to a planning permission. (This valuable provision is similar to Section 11 of the 1977 Pollution Act regarding breaches of Sections 3/4 of that Act.)

Discharges to aquatic environments

Agricultural effluents. The greatest threat to our inland waters in recent years is from agricultural effluents, owing mainly to intensification and new methods of production. Silage-making operations, intensive use of fertilisers, chemical dressings, herbicides and insecticides are causing adverse effects, particularly where there is little control of siting and construction of buildings. The use of land for agriculture and forestry is exempted development under the Planning Acts. There is a measure of control over development of buildings used for housing livestock and poultry. Planning permission is required where such development exceeds 300 square metres in floor area with ancillary provision for effluent storage.

The *entry* of agricultural effluents into waters may be an offence under S.3(1) of the 1977 Pollution Act. They are not licensable under S.4 or S.16, unless they enter waters or sewers via "any works, apparatus, plant or drainage pipe" used for their disposal to waters or to a sewer. Since the cost of treating animal manures or silage effluents to licensable standards would be prohibitively high, it is unlikely that licences will be issued under Ss.4/16. Local authorities are making increasing use of Section 12 notices requiring specified steps to be taken to prevent polluting matter entering waters from premises such as silos, livestock buildings, slurry tanks, etc. Government grant aid schemes are subject to certain measures aimed at safeguarding against pollution but these are of little legal impact.

The unsightly but not uncommon occurrence of decomposing animal carcases in rivers poses problems, particularly after floods. S.48 of the Diseases of Animals Act 1966 prohibits (under penalty up to £100) throwing or placing the carcase of an animal or bird which has died of disease in any river, stream or other water. However, S.3 of the Pollution Act 1977 may be a more effective remedy if the offender can be traced.

Products. There is no overall system for the control of environmentally harmful chemicals and chemical products used in industry, commerce and domestically. Some agrichemicals are controlled by the Departments of Agriculture and Health, mainly by restrictions on distribution in compliance with EEC directives.

The sale of some weedkillers such as Paraquat is regulated under the Poisons Act 1961. The transport of many chemicals is regulated under the Dangerous Substances Act 1972 (see Regulations 1980).

Waste disposal. Various E.C. Waste Regulations 1979/84 charge local authorities with responsibility for supervision of disposal of waste in their areas. Included are general wastes, toxic and dangerous wastes, waste oils, and P.C.B.s, and the Litter Act 1982 also applies.

Exemptions. It should be noted that certain bodies and their operations are exempted from compliance with the Fisheries Acts in respect of water pollution under the following statutes. However, such operations must have due regard for offsetting any damage to fisheries:

> Shannon Electricity Act 1925
> Liffey Reservoir Act 1936
> Arterial Drainage Act 1945
> Electricity Supply (Amdt.) Act 1945
> Turf Development Act 1946
> Local Authorities Works Act 1949

APPENDIX 1
Local Government (Water Pollution)
(Amendment) Bill, 1989
Explanatory and Financial Memorandum

The object of the Bill is to modify and extend the present arrangements for control of water pollution with a view to ensuring that the quality of water resources is maintained at a high level. For this purpose, the Bill amends and extends the Local Government (Water Pollution) Act, 1977, and (in so far as it relates to water pollution) the Fisheries (Consolidation) Act, 1959. In particular, it provides for:

(a) an amendment to the "good defence" provision in Section 3 (3) of the 1977 Act to put a greater onus on the person charged to prove that he could not reasonably have foreseen that his act or omission might cause pollution of waters (Section 3);

(b) the charging by local and sanitary authorities and An Bord Pleanala of fees for licence applications, reviews and appeals (Sections 4, 5, 13 and 14);

(c) the review by local and sanitary authorities of discharge licences at any time where there are reasonable grounds for believing that any of the beneficial uses of the receiving waters is or is likely to be threatened (Sections 5 and 13);

(d)   the extension to any person of the right to apply to the courts for an order seeking the mitigation or remedying of the effects of pollution, and the making of explicit provision for remedial measures such as the replacement of fish stocks and the making good of consequential losses incurred by any person as a result of the pollution (Section 7);

(e)   extending the grounds on which a high court order may be sought to cover situations involving a risk of water pollution (section 8);

(f)   clarification of a local authority's power to serve notices regulating practices (such as silage making and slurry spreading) which, in its opinion, could result in water pollution (Section 9);

(g)   direct intervention by local and sanitary authorities to prevent or deal with any water pollution incidents (Section 10);

(h)   conferring on local and sanitary authorities power to obtain information on water abstractions, discharges to waters and activities or practices which are relevant to their pollution prevention functions (Section 17);

(i)   imposition of a civil liability on polluters for injury, loss or damage suffered by any other person (Section 20);

(j)   new powers for local authorities to make bye-laws regulating or prohibiting specified agricultural activities in their functional areas, or any part thereof, where they consider it necessary to do so in order to eliminate or prevent water pollution; the bye-laws may be appealed to the Minister, who may confirm or annul them or direct that they be amended in a specified manner (Section 21);

(k)   an increase to £1,000 in the maximum fine on summary conviction under the Local Government (Water Pollution) Act, 1977 (from £250), and the Fisheries (Consolidation) Act, 1959 (from £500), and to £25,000 (from £5,000) and/or imprisonment for a period of up to 5 years (from 2 years) on conviction on indictment under the 1977 Act; indictable offences with the same penalties will be available under the Fisheries Act and provision is being made for a "good defence" in respect of prosecutions taken under section 171 of that Act (Sections 24 and 25);

(l)   the payment of fines arising from prosecutions taken by local and sanitary authorities to the authority concerned (Section 26);

(m)   court orders requiring pollution offenders to recoup costs and expenses incurred on the investigation, detection and prosecution of offences (Section 28); and

(n)   the repeal of the provisions of the 1977 Act which provide for the establishment, appointment and functions of the Water Pollution Advisory Council and other appropriate repeals (Section 30).

Section 30 provides for the repeal of certain provisions of the 1977 Act. Sections 3 (4), 4 (9), 6 (5), 12 (6), 14 (3), 16 (9), 19 (5), 27 (4), 28 (8) and 31 are being repealed as a consequence of section 27 of the Bill. Section 2, which is also being repealed, concerns the appointment, composition and functions of the Water Pollution Advisory Council. Section 34 (c), which provided for the repeal of sections 171 and 172 of the Fisheries (Consolidation) Act, 1959, was never brought into operation and the sections have continued to be used by regional fisheries boards when taking prosecutions for water pollution offences. *The repeal of Section 34 (c) will ensure that Sections 171 and 172 may continue to be used by fisheries boards.* A consequential change arising from the retention of these sections involves the repeal of Section 3 (6).

Aidan Barry (S.W.R.F.B., Macroom, Co. Cork): If the water and fish of rivers cannot be privately owned, has the State a right to use those waters by virtue of its ownership, i.e. has the owner of fishing rights sufficient case to prohibit the State using the centre of channel for boating, fishing from boats, etc.?

Sean McMorrow: The *right to fish* in inland (non-tidal) rivers is usually privately owned by the riparian landowner or his grantee. The State will only have such rights where they have been acquired by Commission under various Land Acts. The fishery owner may control the fishing in his stretch of river but he may not prevent navigation as such, which is a right of way over water. The State has no right to use boats on inland waters unless acquired by statute as in certain canals, etc.

John Flynn (Carnagh, Foulksmills, Co.Wexford): Why should County Councils monitor pollution, including their own? Fishery Boards would do the job if given the staff and finance. Their staff are on the rivers and streams on a constant basis. County Councils are the worst pollutors, i.e. raw sewage, etc.

Sean McMorrow: Overall control of water resources, pollution prevention etc. is a matter for national policy. There is a strong case for having an independent water authority in this country. Certainly the Fishery Boards are in a unique position to detect and prosecute pollution offences from their on-the-spot experience and expertise.

In: Steer, M.W. (ed.) 1991 *Irish Rivers : Biology and Management*, pp 279-286. Royal Irish Academy, Dublin.

IRISH RIVERS - THEIR ECONOMIC BENEFITS AND POTENTIAL

Brendan J. Whelan

Economic and Social Research Institute, Dublin

INTRODUCTION

Rivers can have numerous economic benefits: for travel and transport, power generation, drainage, water supply to agriculture, effluent disposal, recreation, etc. In this paper I will concentrate on the economic impact of angling in rivers, how this might be measured and what policies are needed to maximise the potential economic value of river angling. At the ESRI we have carried out several studies relating to this topic over the last number of years (O'Connor and Whelan 1973; O'Connor *et al.* 1974; Whelan *et al.* 1975; Whelan and Marsh 1988).

The paper begins by reviewing briefly the techniques which have been devised by economists to measure the economic impact of angling. It goes on to present some data on the economics of river angling from our recent survey and to contrast these data with similar information relating to lake and sea angling. It concludes with some suggestions regarding policy for the development of river angling in Ireland.

Techniques of economic assessment

For many years now, economists have been wrestling with the problems of conceptualisation and measurement posed by attempts to value angling resources. Whelan and Marsh (1988, 10-13) present a brief synopsis, while Rosen (1974) gives a more comprehensive review. Basically two approaches have been developed.

The first attempts to "value" the angling amenity directly by assessing not just the expenditure incurred on it but also the unpaid-for benefits that often accrue to participants. A number of different techniques have been suggested for measuring these benefits, but most involve ways of estimating the "consumer surplus" as illustrated in Fig. 1. The downward-sloping line is a conventional demand curve, i.e. it represents the quantity of a good which will be purchased (in this case, the number of days fished) at different prices.

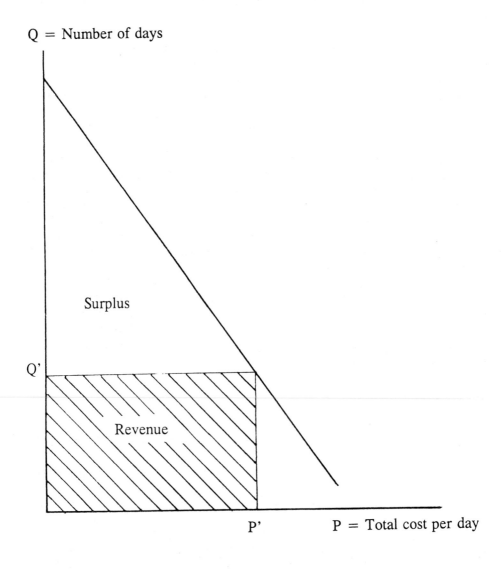

Fig. 1. Hypothetical demand curve for days fished.

Price in this instance includes all the costs associated with a day's fishing (travel, accommodation, fees, tackle, etc.). The line slopes downwards to the right indicating that the higher the price, the lower the quantities purchased (i.e. the fewer days will be fished). The total revenue generated by angling can be measured by the shaded rectangle (= Q' days at price of P'), while the triangle labelled surplus denotes the additional, unpaid-for benefits that accrue to the participants. Clawson and Knetsch (1966) and Davis (1964) are mainly concerned with how this surplus can be estimated.

This approach is mainly used to value angling resources which are heavily utilised by domestic (local) anglers and is concerned that the angling resource be comprehensively valued. It can be difficult to convince policymakers that the estimates derived from it are "real" since the values quoted do not correspond entirely to actual money transactions. The second approach, which is the one usually adopted in the ESRI studies, is more pragmatic. It concentrates on estimating the total gross value generated by angling activity. In statistical terminology this is equivalent to the "gross output" of the angling industry. This concept is of interest because angling is seen as creating economic activity and generating employment that would not otherwise exist. It is assumed that the country's resources, in the form of labour, capital and angling waters, are not being fully utilised. Given the depressed nature of some of the regions with the best angling waters, this seems a reasonable assumption.

The usual methods used to measure gross value involve carrying out sample surveys of the anglers in question so that their total expenditure can be estimated. The Whelan and Marsh (1988) study reports on the results of two such surveys, one of out-of-state (visiting) anglers and the other of Irish anglers. Random samples of both groups were interviewed and data collected on their fishing patterns, opinions, catches and expenditure.

The overall picture

Table 1 shows the estimated numbers of anglers in Ireland, classified by origin and the type of angling they prefer. Irish anglers are estimated to number about 122,000 while there are about 54,000 visiting specialist anglers. Bord Fáilte estimate that a further 126,000 visitors do some angling while on holiday in Ireland but are not specialists (Bord Fáilte Éireann 1987).

Table 1. Estimated numbers of visiting anglers (in 1986) and Irish anglers (in 1987).

| | Type of angler | | | |
	Game	Coarse	Sea	Total
Specialist visitors	15,000	34,000	5,000	54,000
Irish anglers	63,000	16,000	44,000	122,000
Non-specialist visitors				126,000

Table 2 shows total expenditure by all these anglers. Total expenditure by specialists amounts to about £57 million, just over half of which is incurred by visiting anglers. Irish anglers spend about £11 million on day trips, £11 million on overnight trips and £7 million on other current costs.

Table 2. Estimated total expenditure (£m) by anglers, classified by type of angler.

| | Type of angler | | | |
	Game	Coarse	Sea	Total
Irish anglers:				
Day trips	6.36	1.20	3.76	11.31
Overnight trips	5.73	0.43	4.64	10.80
Other current costs	3.46	1.46	2.22	7.14
Total expenditure	15.55	3.09	10.62	29.25
Visiting specialist anglers	12.73	12.89	2.37	27.99
All angling expenditure	28.28	15.98	12.99	57.24

Expenditure attributable to rivers

In this section, I concentrate on the economic effects of angling in rivers, and contrast this with the activity attributable to the two other main locations used by anglers, the sea and lakes. Table 3 shows the breakdown of expenditure associated with angling in these three locations. (Note that this table excludes expenditure on tackle and other current costs.) It can be seen that of the total expenditure of £50 million about one-fifth is generated by sea angling, with the balance being evenly divided between rivers and lakes. Thus, Irish rivers generate about £20 million of angling expenditure each year.

Table 3. Estimated expenditure (£m) on different types of water.

| Type of water | Irish anglers | | | |
	Day	Overnight	Visitors	Total
River	5.2	2.6	12.4	20.2
Lake	2.6	2.8	14.0	19.4
Sea	3.5	5.4	1.6	10.5

Table 4 shows the visiting anglers' overall evaluation of the three types of location. They were asked to rate each location on a scale running from Excellent, through Good and Fair to Poor. In terms of overall evaluation, rivers rate midway between lakes (which were in general judged to be slightly superior) and the sea (which was rated somewhat lower). Among the reasons given by those anglers who gave "fair" or "poor" ratings were over-exploitation of salmon by commercial fisheries, pollution and declines in fish stocks.

Table 4. Rating (%) of different types of waters by visitors.

Rating

Type of water	Excellent	Good	Fair	Poor	Total
River	9.4	26.6	36.8	27.2	100.0
Lake	13.9	29.1	28.1	28.9	100.0
Sea	6.5	16.5	29.1	37.9	100.0

The potential of Irish rivers

Provided that they are conserved and developed, Irish rivers have substantial potential. This view is based on several pieces of evidence. In the first place, the British National Angling Survey (NOP 1970) and the more recent survey by Millward Brown (1987) both show evidence of substantial demand for high-quality angling from British anglers. A similar pattern probably exists in continental Europe where angling waters, particularly rivers, have been depleted by pollution and other factors associated with economic expansion and population pressure. Irish rivers can offer fishermen from these countries some unique experiences since they can fish for natural (not stocked) fish in clean waters. Furthermore, Ireland offers these anglers the chance to "have a go" and fish for species that might not be available (or accessible to them) in their home countries. For instance, salmon fishing in Ireland is relatively inexpensive compared to other countries and would probably be within the means of most visiting coarse anglers.

Further evidence of strong demand for Irish angling comes from our survey data. Table 5 is based on some work being jointly conducted with Dr F. Bell of Florida University (Bell and Whelan in prep.) It shows a regression equation in which we try to predict the numbers of days which an angler will spend in Ireland. The key independent variable of interest here is DMQ, which shows that the effect of having experienced success is both positive and significant. Thus, the more the quality of angling in Irish rivers improves the more visitors it is likely to attract.

Table 5. Least-squares cross-section regression of the demand model
testing main quarry hypothesis: foreign salmon anglers
visiting the Irish Republic, 1986.

(Dependent variable: fishing days per trip)* - (Sample size: 55)

Independent variables		Logarithmic form**	Linear form**	Arithmetic mean of variables IR £
Constant		1.839	-3.812	
		(8.192)	(-1.268)	
TRC	Total basic costs	.166	.004	312.77
	(travel, etc.)	(3.634)	(1.905)	
ONSCD	On site costs (accomm.,	-.180	-.008	58.65
	fishing fees, etc.)	(-2.798)	(-.887)	
DMQ	Did respondent catch	.203	1.571	.691
	main quarry?	(2.011)	(1.489)	
DF1	Proportion of holiday	.705	11.568	.779
	spent fishing	(6.323)	(4.069)	
DF3	Is respondent a fishing	.227	2.265	.509
	club member?	(2.489)	(2.327)	
Y	Annual income	.043	.201	19,675.00
		(.311)	(.502)	
R2		.557	.319	
F		12.302	5.213	

Notes: * Arithmetic mean of DFS = 9.055 days;
geometric mean of DFS = 8.085 days.
** t-values in parentheses.

Maximising the potential

Even this brief analysis of the economic potential of angling in Irish rivers shows that this activity is of substantial economic significance and capable of being further developed. In order to achieve such development several
sets of measures are required. These might be summarised under the headings of conservation, management and marketing.

Conservation. This is probably the most important area needing attention. It
covers such steps as pollution reduction and elimination. Local authorities
in particular have a poor record in caring for the quality of water in rivers,
especially in view of their legal position as the main enforcers of anti-pollu-
tion legislation. Intensive agriculture and industry also pose serious problems which urgently need to be tackled.

Arterial drainage schemes have in the past caused severe deterioration of
river angling and the value of any angling lost should feature prominently in

pre-scheme cost/benefit analyses. Bruton and Convery (1982) shows that many such schemes were of doubtful economic value even before the deterioration or disruption of angling is taken into account. The minimisation of such problems by careful engineering design and rehabilitation should be a key feature of any drainage schemes which are put in place.

Management. Our survey showed that many visitors found access to waters difficult and some complained about lack of information. This suggests that there should be more investment in improving access (stiles, stands, etc.) and additional detailed information provided about the availability and nature of angling in Ireland.

Another area mentioned by many of the salmon anglers we interviewed related to what they saw as excessive exploitation of salmon by commercial netsmen. This issue has been widely discussed (e.g. Salmon Review Group 1987; Whelan *et al.* 1975; Whelan and Whelan (1986). The latter paper showed that in recent years the argument in favour of utilising the natural salmon stocks for angling rather than commercial exploitation had strengthened considerably.

Marketing. I would like to conclude with a personal view, the evidence for which is difficult to quantify. It is my impression that the promotion of Irish freshwater angling emphasises lakes at the expense of rivers. Given the dearth of clean rivers in the rest of Europe and the heavy usage of "put and take" lakes in many places there, it would seem that Ireland has a very strong comparative advantage in providing clean rivers with good natural stocks of many species. The distinctive aspects of Irish river angling should, therefore, be underlined in all promotional activities.

REFERENCES

BELL, F.W. and WHELAN B.J. (in prep.) The Main Quarry Hypothesis and Salmon Angling in Ireland.

BORD FÁILTE ÉIREANN 1987 *A Strategic Brand Marketing Plan for Irish Angling.* Dublin. Bord Fáilte.

BRUTON, R. and CONVERY F. 1982 Land drainage policy. In *Ireland.* Economic and Social Research Institute, Policy Research Series, No. 4. Dublin.

CLAWSON, M. and KNETSCH, J.L. 1966 *Economics of Outdoor Recreation.* Baltimore, Maryland. Johns Hopkins University Press.

DAVIS, R.K. 1964 *The Value of Big Game Hunting in a Private Forest.* Transactions of the North American Wildlife and Natural Resources Conference, Vol. 29, 393-403.

MILLWARD BROWN LTD 1987 *National Angling Survey.* London. National Anglers' Council.

NATIONAL OPINION POLLS 1970 *National Angling Survey.* London. National Environmental Research Council.

O'CONNOR, R. and WHELAN, B.J. 1973 *An Economic Evaluation of Irish Salmon Fishing, Part I: The Visiting Anglers.* Economic and Social Research Institute, General Research Series, No. 68. Dublin.

O'CONNOR, R., WHELAN, B.J. and McCASHIN, A. 1974 *An Economic Evaluation of Irish Salmon Fishing, Part II: The Irish Anglers.* Economic and Social Research Institute, General Research Series, No. 75. Dublin.

ROSEN, S. 1974 Hedonic prices and implicit markets: product differentiation in pure competition. *Journal of Political Economy,* 82, 34-55.

SALMON REVIEW GROUP 1987 *Report: Framework for the Development of Ireland's Salmon Fisheries.* Dublin. Stationery Office.

WHELAN, B.J. and MARSH, G. 1988 *An Economic Evaluation of Irish Angling.* Dublin. Central Fisheries Board.

WHELAN, B.J. and WHELAN, K.F. 1986 *The Economics of Salmon Fishing in the Republic of Ireland.* Proceedings of the 17th Annual Study Conference of the Institute of Fisheries Management, University of Ulster, Coleraine.

WHELAN, B.J., O'CONNOR, R. and McCASHIN, A. 1975 *An Economic Evaluation of Irish Salmon Fishing, Part III: The Commercial Fishermen.* Economic and Social Research Institute, General Research Series, No.78. Dublin.

POSTER ABSTRACT

THE VEGETATION OF IRISH RIVERS

Hester Heuff

Blackrock, Co. Dublin

Fifty-six river stretches, on 32 rivers, were examined botanically. The study sites were located throughout the country in catchments representing each of six main geological types. Fourteen major ecological niche types were identified and 39 vegetation units were described, using the Braun-Blanquet method. Several of these vegetation types are thought to be new to science. Eighteen of the vegetation units were used to classify the river stretches, giving rise to six different stretch types. Several of these types are almost certainly very rare in Europe and most are threatened by eutrophication and/or drainage. It is recommended that a programme of active nature conservation of rivers be initiated, so that a full range of Irish river types may be conserved, either through a nature reserve approach or, in the majority of cases, through planning control and management arrangements. The latter will require cooperation between all organisations with responsibility in these areas, including planning authorities, drainage boards, inland fisheries, wildlife service etc.

This work was carried out under contract to the Wildlife Service, Office of Public Works.